THE PRINCIPLES AND PRACTICE OF RADIO DIRECTION FINDING

THE PRINCIPLES AND PRACTICE OF RADIO DIRECTION FINDING

BY

CHARLES H. COTTER

Ex.C., B.Sc., M.I.N.

LONDON

SIR ISAAC PITMAN & SONS, LTD.

First published 1961

SIR ISAAC PITMAN & SONS, LTD.
PITMAN HOUSE, PARKER STREET, KINGSWAY, LONDON, W.C.2
THE PITMAN PRESS, BATH
PITMAN HOUSE, BOUVERIE STREET, CARLTON, MELBOURNE
22–25 BECKETT'S BUILDINGS, PRESIDENT STREET, JOHANNESBURG

ASSOCIATED COMPANIES
PITMAN MEDICAL PUBLISHING COMPANY, LTD.
39 PARKER STREET, LONDON, W.C.2
PITMAN PUBLISHING CORPORATION
2 WEST 45TH STREET, NEW YORK
SIR ISAAC PITMAN & SONS (CANADA), LTD.
(INCORPORATING THE COMMERCIAL TEXT BOOK COMPANY)
PITMAN HOUSE, 381–383 CHURCH STREET, TORONTO

MADE IN GREAT BRITAIN AT THE PITMAN PRESS, BATH
F1—(T.936)

PREFACE

THIS small book is intended primarily for the use of navigating officers of the Merchant Navy, especially those students of navigation who are preparing for either the First Mate or Master examinations set by the Ministry of Transport and Civil Aviation.

Candidates for the First Mate examination are examined on their ability to operate a radio direction finding equipment, and are required to have a knowledge of the practical use of the instrument as used on board a ship. Candidates for the Master examination are required to have some knowledge of the principles of radio direction finding, and accordingly relevant questions are set in the examination paper on Navigational Aids. It is to be hoped that both classes of students, as well as qualified and practical navigators, will find the following notes of value. It cannot be overstressed that complete knowledge of any instrument of navigation can be acquired only if the instrument is used frequently for practical purposes under practical conditions. This applies particularly to the radio direction finder: no amount of book-learning can substitute for experience at operating it.

One, if not the most important, of the objectives of a navigator, in respect of any instrumental aid to navigation that he may have occasion to use, is for him to be able to assess the degree of accuracy —and hence the value—of the particular instrument under any set of the varied conditions in which it may be used. The inability of many navigators to do this for the radio direction finder is doubtless one of the reasons resulting in the instrument being regarded by some as anything but a satisfactory aid to navigation. It is my contention that the more knowledge a navigator acquires about the way in which any of his instruments functions, and the more practice he has at using that instrument, the better he is able to estimate the degree of reliability to place upon the results of observations determined from its use, and accordingly the more valuable to him will be the instrument as an aid to safe navigation.

Most deep sea ships are required by law to be equipped with an efficient radio direction finder. In many such ships the instrument is operated by the Radio Officers of the ship. These officers doubtless have a firm knowledge of the radio principles of the instrument, but they may lack knowledge of certain navigational principles, without the application of which the results of radio direction finding may be rendered incomplete and therefore valueless. In some ships the navigating officers have access to the radio direction finder, and, if this is the case (as it ought to be), advantages should be taken of using it, not only when occasion demands its use for purposes of navigation, but also in order to acquire the necessary practice without which results of observations may be useless. In all cases, co-operation between navigating officers and the radio department of the ship is desirable and essential—not simply to render it unnecessary for each department to have a knowledge of the other's technical field, but to facilitate the specialized knowledge of both being brought together and used for the common good—in this case, the safe navigation of their ship.

The navigator will realize that a full knowledge of the electronics and radio involved in wireless direction finding is outside his horizon—at least at the present time. His training does not provide for it, and neither his employer nor his examiner demands it. It is not unreasonable however for a navigator to be expected to have sufficient knowledge of the instruments, with which he may be provided, to be able to use them intelligently. Sad to say that the old-fashioned maxim "kick it if it doesn't work" still applies in some quarters.

The book is divided into four main sections. Part 1 contains notes on electricity, electronics, and radio, sufficient it is hoped for an understanding of the basic principles of radio direction finding; Part 2 deals with these principles; and Part 3 treats of the practical uses of radio direction finding in merchant ships. The notes on Consol in Part 4 are worthy of the attention of practical navigators.

For some years I have had the pleasure of assisting prospective candidates for the senior grades of the M.O.T.C.A. examinations, at the Sir John Cass College, London. The notes used in lectures on Navigational Aids given to these students have formed the nucleus of the following text. My notes, as might be expected, are based on practical experience at using a direction finder at sea, as well as on many books and technical papers on the subject. Despite the numerous textbooks which deal with radio direction finding, there is none in my opinion which is quite suitable for the men to whom this small book is addressed: this provides one reason for writing. Again, it has been said that every man owes something to his profession, and accordingly, as a curious blend of sailor and teacher, I launch this small work in the hope that it will be of value and interest to practical navigators, and at the same time meet with the approval of their instructors. This is my main reason for writing.

C. H. COTTER

'Stoneden'
Hawkhurst.
August 1959

ACKNOWLEDGMENTS

THE author has pleasure in thanking The Marconi International Marine Communication Company Limited, Siemens, Edison Swan Limited, and The International Marine Radio Company Limited, for their willingness to allow him to use material from instruction manuals and pamphlets pertaining to the direction finders manufactured by these companies.

Permission to use extracts from the following list of papers printed in the journals of the Institute of Navigation is gratefully acknowledged—

"Radio Aids to Navigation," by Sir R. Watson-Watt.
"Consol," by A. H. Jessell.
"The Range and Accuracy of Consol," by A. H. Jessell.
"Marine Radio Position Fixing Systems," by H. E. Hogben.
"The Use of D.F. at Sea," by F. P. Best.
"The Requirements for Radio Aids at Sea," by F. J. Wylie.

It is the author's pleasant duty to record his thanks to his colleague E. J. Doherty, Esq. for reading through the original manuscript, and the courtesy of the Controller of Her Majesty's Stationery Office for allowing the use of extracts from the Admiralty List of Radio Signals, Volume 2. The M.O.T.C.A. Notice M.402 is also acknowledged with thanks.

Finally, thanks are due to the Technical Editor of the publishers and his staff, who have, with great care, produced a neat little book which it is to be hoped will be found to be valuable in the hands of those students of navigation for whom it has been written.

CHARLES H. COTTER

CONTENTS

Part 1

General Electricity, Electronics, and Radio

The term *matter* denotes any material thing which occupies space. Matter is broadly divided into a two-fold division of *elements* and *compounds*. Elements include only those materials which are composed of *atoms* of the same kind, whereas compounds include those materials which are composed of a mixture of atoms. Water, for example, is a compound, as it is a mixture of the elements hydrogen and oxygen in the ratio two to one. The chemical symbol for water is H_2O, and the smallest particle of water, that is to say, a particle formed by two atoms of hydrogen and one atom of oxygen, is known as a *molecule* of water.

An atom consists of one or more (up to about ninety) *electrons*, which revolve at high speed around a central nucleus composed of *protons* and electrons. Electrons are considered to be negatively-charged electrically, whereas protons are positively-charged. An atom in a normal, or stable, state is electrically neutral, that is to say, the number of electrons it possesses is equal to the number of its protons.

The circulating electrons of atoms are interchangeable. Frequently the outermost orbits of circulating electrons of neighbouring atoms cross, giving rise to the possibility of an electron collision which may result in an electron becoming detached from its parent atom. When this happens, the atom which has been deprived of an electron becomes positively-charged, as it now has a surplus of protons. In this state, a positively-charged atom soon acquires another electron, because of its ability to attract electrons.

Electron Flow

At any instant of time, there is in any material a certain number of electrons which have been detached from atoms. These are known as *free* electrons, and they flow randomly from one positively-charged atom to another positively-charged atom. Now if free electrons can be marshalled in some way, and forced to drift in one direction only, they may be made to perform useful work, and the energy associated with such a flow is known as *electrical energy*. The organized flow of electrons with which electrical energy is associated is known as *electricity*.

Electrons will flow, when forced to do so, more readily in some substances than in others. Substances in which electrons may flow with relative ease are said to be good *conductors* or bad *insulators*. There is no material in which electrons cannot be made to flow, and yet there is no material in which electrons may flow with perfect ease: the terms conductor and insulator are therefore relative terms. In a good conductor there is a relatively large number of free electrons, and therefore it is relatively easy to cause a flow of electrons in a good conductor. In a poor conductor the reverse is the case. The force necessary to produce electricity is known as an electron-moving or *electromotive force*, abbreviated to e.m.f.

Before electricity may be detected, a large number of electrons must be involved in the flow, so that the unit of quantity of electricity is a very large number of electrons known as the *coulomb*. In fact, one coulomb is $6{\cdot}29 \times 10^{18}$ electrons. The rate of flow of electrons is expressed in *coulombs per second*, and the name given to the unit of rate of flow is the *ampere*, which is a rate of flow of one coulomb per second. The rate of flow of electrons is known as *current*. Thus if the current in a conductor is say I amperes, then it follows that I coulombs pass any fixed cross-sectional area of the conductor in every second. The unit of quantity of electricity may be defined as the *ampere-second* because, if the current is one ampere, the quantity passing any fixed cross-sectional area of the conductor in one second is one coulomb. A more convenient unit (because more practical) is the *ampere-hour*, this being, by the same argument, 3,600 coulombs.

Electrical Energy

It was mentioned earlier that the energy associated with an electron flow is known as electrical energy. Now energy is simply ability to do work, and is manifested in a variety of forms. Energy may be converted from one form into another, and the production of an e.m.f.—in order to make electrons flow—involves the transformation of chemical, mechanical, or some other form of energy into electrical energy. Just as any other form of energy may be converted into electrical energy, so electrical energy may be converted into any other form.

The three most important effects of an electric current are heating, magnetic, and chemical effects, these being manifested when electrical energy is actively being converted into heat energy, magnetic energy, or chemical energy respectively.

Work is said to be done whenever a force acts through a distance, and a *force* is described as something which when applied to a mass causes it to move, if the mass is free to move, at an ever-increasing velocity. In other words, when a force is applied to a mass which is free to move the mass accelerates. The unit of force in the metric system is the *dyne*. This is the force which when acting on a mass of one gramme causes it to accelerate, if free to do so, at the rate of one centimetre per second per second. If a force of one dyne acts on a mass of one gramme, then, for each centimetre of movement of the unit mass, a unit of work is performed, and the ability to do this represents a unit of energy. Whenever work is being done, energy is being converted from one form into another. The name given to the unit of energy (or work) in the metric system is the *erg*. The practical unit of electrical energy is known as the *joule*, which is equivalent to 10^7 ergs.

A device, in parts of which some form of energy is being converted into electrical energy, and in other parts of which electrical energy is being converted into some other form of energy, is known as an *electrical circuit*. In that part of an electrical circuit where some form of energy is being converted into electrical energy, there is said to be an e.m.f. acting. The unit of e.m.f. is the *volt*, this being the e.m.f. developed when the equivalent of one joule of energy is converted into electrical energy for each coulomb that passes. That is to say, one volt is equivalent to one joule per coulomb.

Where, between any two points in an electrical circuit, electrical energy is being converted into some other form of energy, a *potential difference* or p.d. is said to exist. The unit of p.d., like that of e.m.f., is the *volt*, which is that p.d. existing between two points in a circuit where one unit of electrical energy has been changed to some other form of energy as a consequence of one coulomb passing between the points.

Electrical Potential

When amber is rubbed with silk, it is found that the amber acquires the property of being able to attract small objects. During the rubbing process electrons are transferred from the amber to the silk, and consequently the amber becomes deficient in electrons and the silk acquires a surplus. The amber is said to be positively-charged, and the silk negatively-charged. Now positively-charged (or negatively-charged) bodies tend to repel one another, whereas bodies which have charges of opposite nature tend to attract one another. The force of attraction or repulsion is dependent upon the magnitude of the charges (measured in coulombs) and the distance between the bodies, and is referred to as an *electrostatic force.* For any given charges, the force acting between two charged bodies varies inversely as the square of the distance between the bodies.

No electrostatic forces emanate from uncharged bodies, and such bodies are said to have zero or no *electrical potential.* It is convenient to consider the Earth itself as an uncharged body, and the electrical potential of the Earth therefore is considered to be zero. Electrical potential may be either positive or negative, and electric lines of force, forming an *electrostatic field,* are considered to extend between the two bodies.

If two bodies which have different electrical potentials are connected by means of a conductor, electrons will flow in the conductor until the difference of potential between the two bodies is zero. If a charged body is connected by means of a conductor to the Earth, electrons will flow along the conductor until the potential of the body is zero. The body is said to have been positively-charged if the flow of electrons had been from the Earth to the body, and negatively-charged if the flow had been from the body to the Earth.

If the potentials at any two points in a conductor are different from each other, that is to say, if a p.d. exists between the two points, then a stream of electrons will exist in the conductor so long as the p.d. between the two points is maintained.

Ohm's Law

Now, all conductors to varying degrees resist the flow of electrons, and for any given p.d. between two points in a conductor the current, or rate of flow of electrons, is dependent upon the resisting ability of the conductor, a property known as *electrical resistance.* Likewise, all conductors to varying degrees are capable of allowing electrons to flow in them, the rate of flow being dependent upon a property known as *conductance.*

The relationship between p.d., current, and electrical resistance (or conductance) was first formulated by G. S. Ohm, and is known as *Ohm's law,* which is—

For a constant p.d., current varies inversely as resistance, or directly as conductance, that is

$$V \propto R \times I \qquad \text{or} \qquad V \propto I/M$$

where V is p.d. between two points,
R is the resistance between the two points,
M is the conductance between them, and
I is the current.

The unit of resistance is the *ohm,* which is defined as that resistance offered by a conductor in which the current is one ampere and the p.d. between the ends of which is one volt. Thus, if units are volts, amperes, and ohms, the constant of variation in Ohm's law is unity and so

$$V = R \times I$$

The unit of conductance is the *mho,* which is defined as the conductance of a body which has a resistance of one ohm.

Direction of Current

It was discovered in 1819, by the Danish scientist Oersted, that when a conductor is carrying electricity, the region around the

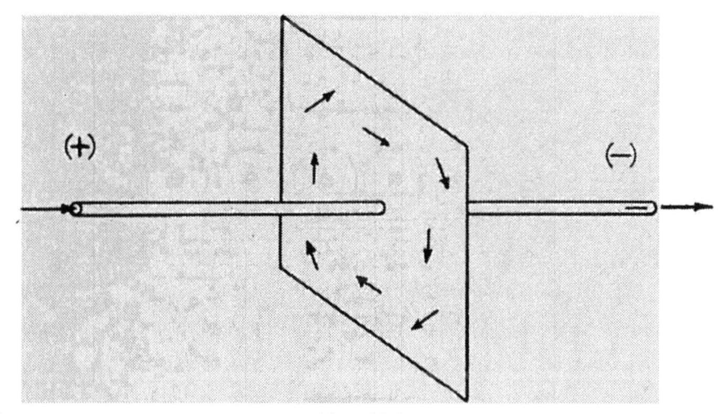

FIG. 1.1

conductor is influenced by the electricity in such a way that the direction of a magnetic needle is affected according to the position of the needle with respect to the conductor, and to the direction of the flow of electrons in the conductor. The effects of electricity on a nearby magnetic needle, stated in a rule known as Ampère's rule, may be given as follows—

Imagine yourself to be swimming in a conductor in the direction of the current with your face towards the needle; the red end of the needle will be deflected towards your left hand.

The direction of the current in this rule is known as the conventional direction, and is in the reverse direction to the electron flow. In Fig. 1.1, which represents the directions taken up by magnetic needles in several positions around a conductor, the conventional direction of the current is indicated by arrows along the conductor. The electron flow is from the end marked − to the end marked +.

If a conductor is wound into the form of a coil, the magnetic effect of electricity in the coil is such that the ends of the coil acquire red and blue polarity, the coil thus developing a magnetic field around it, similar to that around a bar magnet.

Fig. 1.2 illustrates the magnetic effect of electricity in a coil. By applying Ampère's rule it can readily be deduced that the left-hand

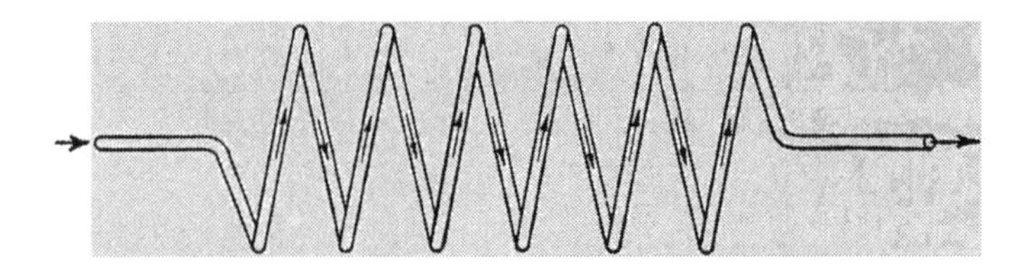

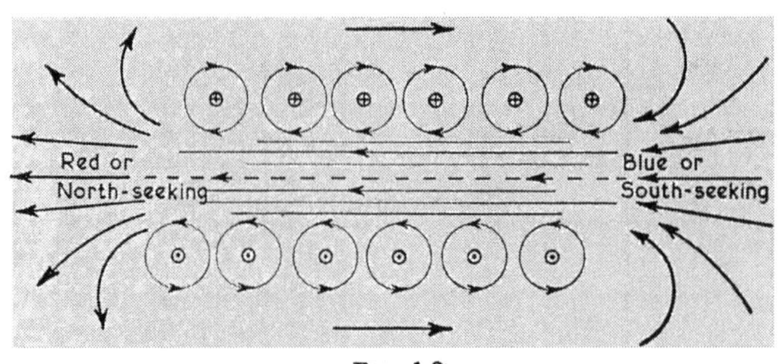

FIG. 1.2

end of the coil in the diagram acquires red polarity, and the other end, blue.

The Electromagnetic Field

The region around a conductor carrying electricity, in which a magnetic needle is influenced, is considered to contain lines of magnetic force forming what is known as an *electromagnetic field.* The energy associated with an electromagnetic field may be harnessed to perform useful work. A most important and useful effect, which is employed in a variety of ways, is one known as the motor effect, which is illustrated in Fig. 1.3.

Suppose a conductor, which is carrying electricity as illustrated in Fig. 1.3, is placed in a magnetic field. The resultant magnetic field of that due to the electric current in the conductor and that due to the field magnets forces the conductor to move in a direction such that the direction of current (conventional), the direction of the magnetic field (from red to blue polarity), and the direction of

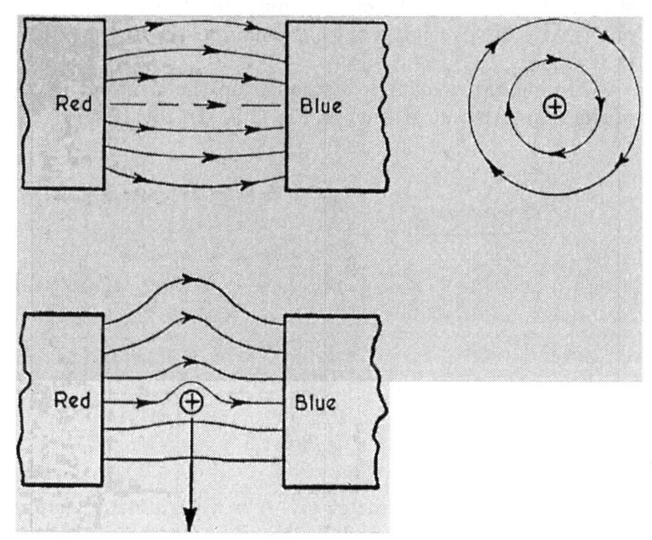

FIG. 1.3

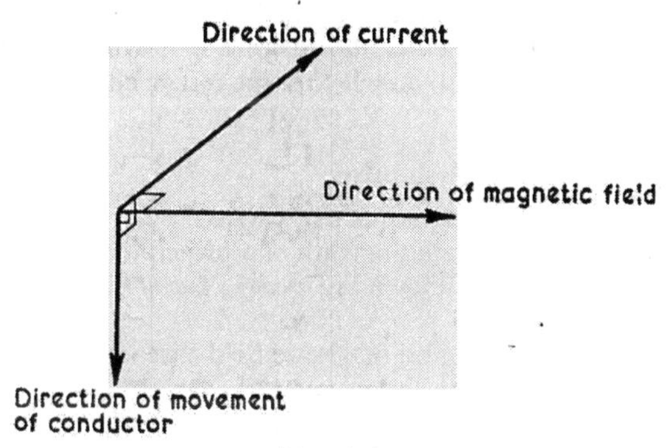

FIG. 1.4

the movement of the conductor are all mutually at right-angles, as illustrated in Fig. 1.4.

The effect described above is employed in the electric motor and also in the action of certain electrical measuring instruments, such as the galvanometer and ammeter, for measuring current, and the voltmeter, which is an instrument for measuring the p.d. between any two points in an electrical circuit.

The motor effect is one in which electrical energy is converted into a form of mechanical energy manifested by the motion of the moving conductor. An electric motor is often described as a machine in which mechanical energy is produced from electrical energy. The scientist Michael Faraday is credited with being the first to demonstrate that mechanical energy may be converted into electrical energy, by discovering what has become known as the *dynamo effect or principle.*

A simple experiment to demonstrate the dynamo effect consists of the apparatus illustrated in Fig. 1.5. In this diagram B is a coil

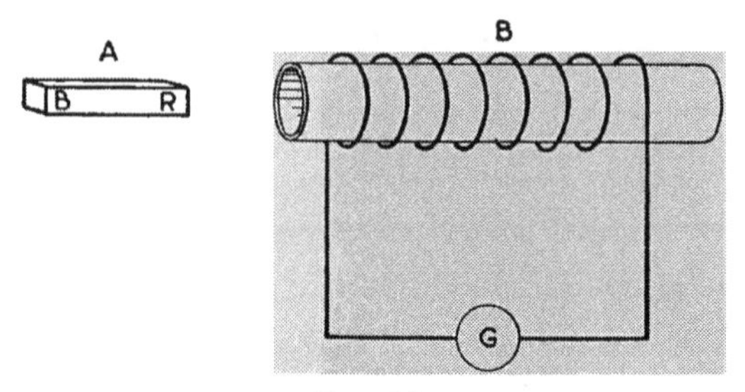

FIG. 1.5

connected to a galvanometer G. A is a permanent magnet which may be thrust into the coil B. Experiment shows that, when the magnet is moving along the axis of the coil, the galvanometer indicates a current in the coil. The direction of the current is dependent upon whether the magnet is moving towards or away from the coil, and also on whether the red or blue end of the magnet is nearer to the coil.

Electromagnetic Induction

The results of experiments, carried out with the apparatus described, demonstrate the truth of a principle which is summed up in what has become known as *Faraday's law of electromagnetic induction*, which is—

Whenever a changing magnetic field threads a circuit, a current is found to exist in the circuit.

The current is said to be due to an *induced electromotive force*, and the magnitude of the current is dependent upon the rate at which the magnetic field is changing. It cannot be too strongly emphasized that an induced current exists only when the magnetic field, which causes it, is changing. A constant field threading a circuit, even though it be a very strong field, will not induce current in the circuit which it is threading.

Faraday's classic experiment, in which he demonstrated the dynamo effect, is illustrated in Fig. 1.6. An iron ring R is wound with two windings of insulated wire to form a primary winding P, which is connected to a source of e.m.f. and a switch; and a secondary winding S, which is connected to a galvanometer G.

When the switch in the primary circuit is made or broken, the galvanometer will indicate a momentary current in the secondary

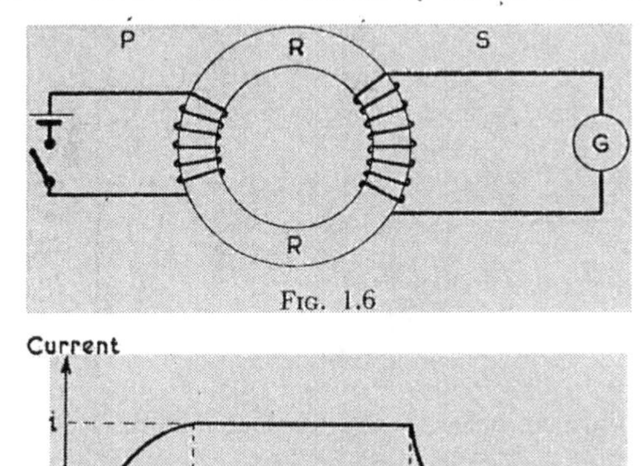

FIG. 1.6

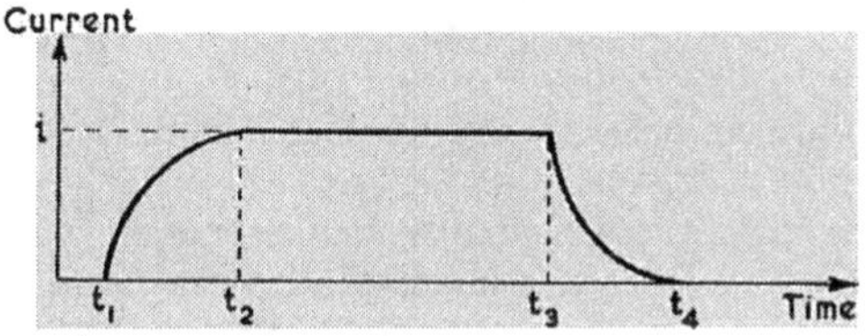

FIG. 1.7

circuit. After the instant the switch in the primary circuit is made (or broken), the magnetic field around the primary coil grows (or decays), for an interval of time, the duration of which depends upon a property of the primary and secondary circuits known as *inductance*.

The graph of current against time in the primary or secondary circuit described above is illustrated in Fig. 1.7. If the switch in a primary circuit is made at the instant t_1, then the current grows and reaches a constant value of i amperes at the time t_2. If the switch is broken at the instant t_3, then the current falls until the instant t_4, at which time it is zero. The interval t_1 to t_2, during which the current increases, is equal to the interval t_3 to t_4, during which the current decreases to zero.

During either the interval t_1 to t_2, or the interval t_3 to t_4, the current in the primary circuit (see Fig. 1.6) is changing, and consequently the magnetic field threading the primary and secondary circuits is also changing. The changing magnetic field threading the secondary winding induces a current in the secondary circuit which is indicated by the galvanometer.

Faraday's experiments showed that, not only is a current induced in a circuit whenever the circuit is threaded by a changing magnetic field, but also that the induced e.m.f. is proportional to the rate of change of magnetic field strength. The direction of an induced current depends upon whether the threading magnetic field is increasing or decreasing in intensity.

An electrical law, which is closely allied to Faraday's law of electromagnetic induction, is Lenz's law, by means of which the

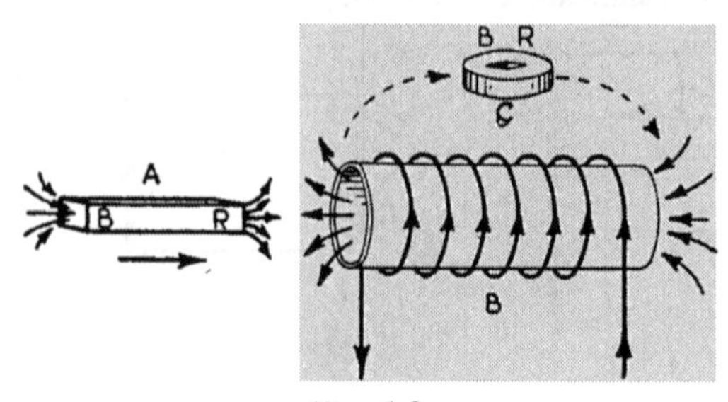

Fig. 1.8

direction of an induced current may be determined. Referring to Faraday's experiment illustrated in Fig. 1.6: in order to generate an induced current in the secondary circuit, energy is required. The energy is derived from the mechanical work done in overcoming the mutual forces of attraction or repulsion which exist between the magnetic fields due to the currents in the primary and secondary circuits. Lenz's law may be stated thus—

An induced current acts in such a direction that its magnetic field acts in opposition to the changing field which causes the induced e.m.f.

To demonstrate Lenz's law, the apparatus illustrated in Fig. 1.8 may be used. Suppose a magnet A is thrust into the core of a coil B, the coil forming part of a closed circuit. The direction of the induced current in coil B may be determined by means of the exploring compass C. When the red end of the magnet is thrust into the coil, as illustrated, the magnetic effect of the induced current in the coil is such that the coil acts as a permanent magnet, with its red end nearer to the red end of the magnet. In other words, the magnetic field that develops around the coil acts in opposition to the field of the magnet, and the direction (conventional) of the induced current is as indicated.

Consider the circuit depicted in Fig. 1.9. After the switch is closed, for a small interval of time, the duration of which depends upon the inductance of the circuit, the current increases, and during this time the magnetic field around the coil is one of increasing intensity. During this interval there are two e.m.f.s acting. One is the cell e.m.f. and the other is the induced e.m.f. which opposes the cell e.m.f. When the switch is broken, the induced e.m.f.—again acting in opposition to the cell e.m.f.—will tend to cause the current to be maintained. This explains why, in circuits of high inductance,

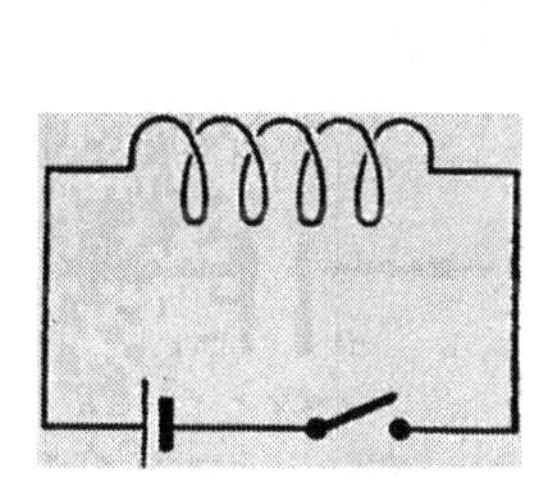

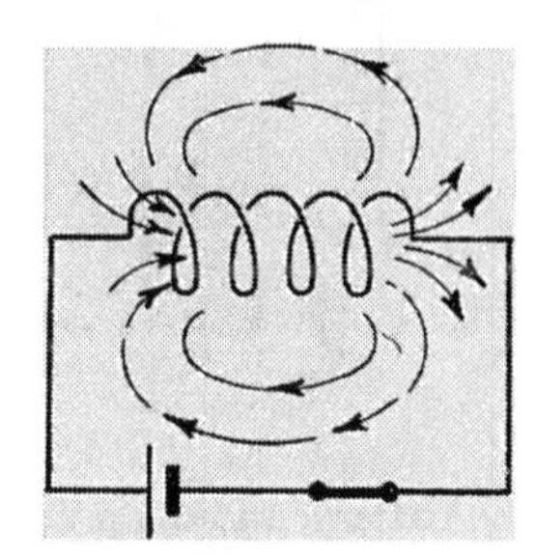

FIG. 1.9

a spark crosses the gap between the points of the switch when the circuit is broken.

Inductance is often regarded as the ability of a circuit to oppose changes in current. The unit of inductance is the *henry*. This is the inductance of a circuit in which the induced e.m.f. is one volt consequent upon a rate of change of current of one ampere per second.

In a circuit in which the induced e.m.f. is E volts consequent upon a rate of change of S amperes per second, the inductance of the circuit is L henrys where

$$L = \frac{E}{S}$$

An e.m.f. may be induced in a circuit if there is a change of current in a neighbouring circuit. A pair of circuits is said to have a *mutual inductance* of one henry if an induced e.m.f. of one volt in one of the circuits is due to a change of current at the rate of one ampere per second in the other.

The Condenser (or Capacitor)

The simplest form of condenser consists of two metal plates separated from one another by an insulating material such as glass, mica, air, etc. If the plates of such a condenser are connected to the terminals of a battery as illustrated in Fig. 1.10(*a*), no circuit is formed and therefore no current exists. If, however, the condenser is disconnected from the battery and connected to a galvanometer as in Fig. 1.10(*b*), the galvanometer will indicate a temporary current.

The plate B of the condenser, which was initially connected to the negative terminal of the battery, acquires a negative charge, whereas plate A acquires a positive charge. The plates remain in the charged condition after the condenser has been disconnected from

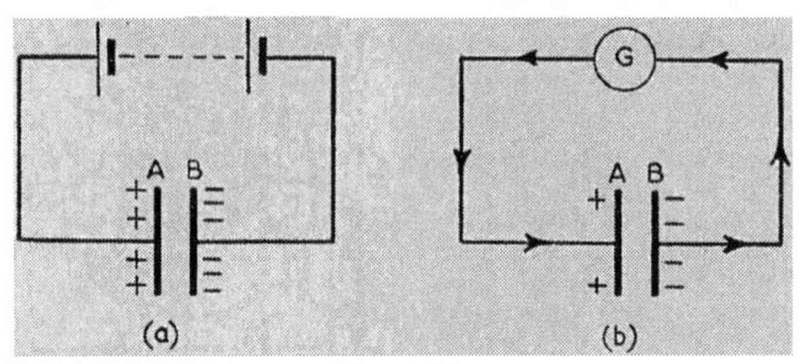

FIG. 1.10

the battery, and the p.d. which exists between the plates results in a current when the plates are reconnected through the galvanometer.

The ability of a condenser to create a flow of electrons, after being charged, is referred to as the capacitance of the condenser. The unit of capacitance is the *farad*. This is the capacitance of a condenser (or a circuit) charged with one coulomb, and the p.d. between the plates is one volt.

If Q is the charge in coulombs, and V is the p.d. between the plates of the condenser in volts, then the capacitance of the condenser in farads is C, where

$$C = \frac{Q}{V}$$

A charged condenser possesses energy in the form of an electrostatic field which exists between the plates. The capacitance of a condenser or a circuit is often regarded as a measure of the ability of the condenser or circuit to store energy in the form of an electrostatic field, whereas the inductance of a circuit is regarded as the ability of a circuit to store energy in the form of an electromagnetic field.

Oscillatory Circuits

If a charged condenser is allowed to discharge through a coil, the charge surges through the coil, and the plates of the condenser become alternately positively and negatively charged. The character of the oscillations set up in such a circuit is of the greatest importance in its application to the radiation and reception of radio energy. Careful study of the oscillatory discharge of a condenser, as it is called, is worthy of the closest attention of the student.

Referring to Fig. 1.11, sketch (*a*) represents a charged condenser,

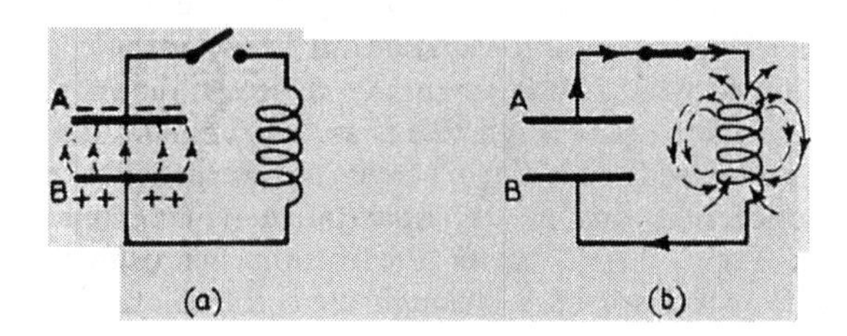

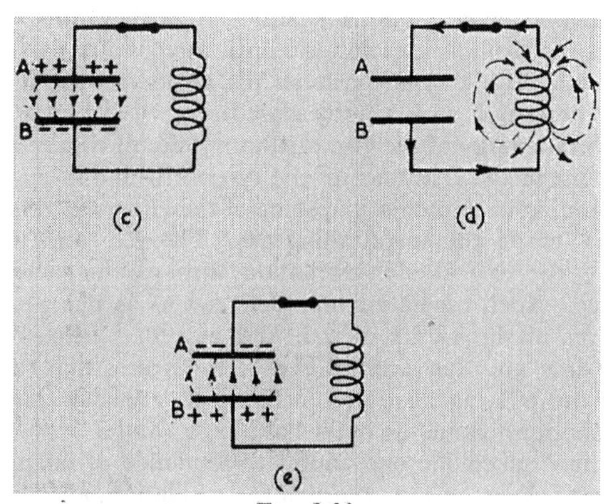

Fig. 1.11

with plate A negatively charged, connected to a coil. After the switch has been closed, a flow of electrons from plate A, through the coil to plate B, results in the condenser discharging. Initially, all the energy in the system was in the form of an electrostatic field between the condenser plates. When current exists in the circuit, however, part, at least, of the energy is electromagnetic, manifested by the magnetic field around the coil. As the electrostatic component of the total energy of the system diminishes, that of the magnetic energy increases. When the p.d. between the plates of the condenser is nil, as illustrated in Fig. 1.11(*b*), the current is maximum, and all the energy of the system is contained in the magnetic field associated with the coil. Notice that the current in the system is maximum when the p.d. is zero. After the instant when the current is maximum, the inductance of the system causes the electrons to continue flowing, with the result that they pile up on plate B of the condenser, and so a p.d. develops as the current decreases. When the p.d. is maximum in magnitude but opposite in direction to what it was originally, the current in the system is nil, and, at the instant when this is so, all the energy of the system is again associated with the electrostatic field between the plates of the condenser, due to its capacitance. This condition is depicted in Fig. 1.11(*c*). After the instant when this occurs, the electrons flow from plate A, through the coil to plate B, and again as depicted in Fig. 1.11(*d*), when the p.d. is nil, the current is maximum and all the energy of the system is due to the inductance of the system. Finally, to complete one oscillation of the charge, the electrons pile up on plate A again, and as they do so the energy becomes increasingly electrostatic until the condition of the system, as shown in Fig. 1.11(*e*), becomes the same as what it was at the instant the switch was initially closed.

The total energy of such an oscillatory circuit diminishes, mainly on account of the resistance of the system, which is responsible for converting, at each oscillation, some of the energy into heat energy, which is lost to the surrounding air. The p.d. and the current, therefore, decrease at each oscillation, until all the energy has been converted. Such oscillations are referred to as being *damped*, and the curves of the values of p.d. and current plotted on a graph against time are illustrated in Fig. 1.12. Notice that the curves of current and p.d., as illustrated in Fig. 1.12, are such that when one has a maximum value the other has a zero value.

The time taken for one complete sequence of changes of p.d. or current is known as the *period* of the oscillation, and the complete sequence—representing a surge of electrons from one plate of the condenser to the other and back to the first plate—is referred to as a *cycle*. Each cycle is divided into 360 degrees. It follows, therefore, that the p.d. and current cycles are separated by 90 degrees, and they are therefore said to be *out of phase* with one another by this angular *phase difference*.

The period of oscillation in the type of circuit described is a very small interval of time. The number of oscillations or cycles completed in one second is normally in the order of tens or hundreds of thousands. The number of cycles made in one second is known as the *frequency* of the oscillations, denoted by f. If the period is T seconds, then

$$f = \frac{1}{T} \text{ cycles per second}$$

The circuit described above, and illustrated in Fig. 1.12, is one in which energy losses are due solely to heating of the circuit due to its resistance. The term *closed oscillatory circuit,* which applies to this

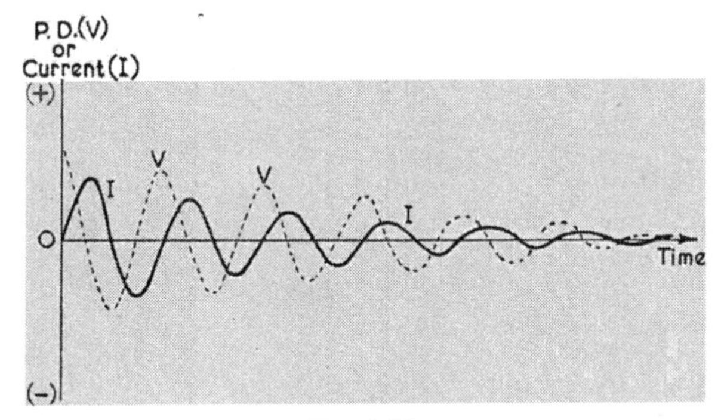

FIG. 1.12

type of circuit, means that no energy associated with the fields of either condenser or coil is lost. If the plates of the condenser are separated so that the electrostatic field is diffused over a large space, the circuit becomes an *open oscillatory circuit,* and energy is lost to the circuit, not only due to heating of the circuit, but also due to field energy being radiated away from the circuit. The radiated energy is of a type known as *electromagnetic energy,* as associated with it are the electric and magnetic fields due to the capacitance and inductance of the circuit respectively.

Radio Waves

The basic principle involved in the production and radiation of wireless or radio energy rests in the oscillatory nature of the discharge of a charged condenser through an inductive circuit. In 1865, James Clerk Maxwell showed by mathematical investigation that the oscillatory discharge of a condenser would result in the propagation of a form of electromagnetic energy which would move outwards from the oscillatory circuit at a velocity of 300 million metres per second. Maxwell also deduced that this type of energy is similar in nature to that of light, radio energy differing from light only in respect of the frequency of the oscillations.

More than two decades were to elapse after Maxwell's brilliant investigation was made, before it was demonstrated that the radio energy described by Maxwell could be transmitted and detected at some distance from the place where it was generated. This extremely important scientific result was achieved by Heinrich Hertz in 1888.

Wireless or radio energy is transmitted or radiated by means of a radio transmitter, which is essentially a device in which a rapidly oscillating motion of electrons is produced: the oscillating electrons

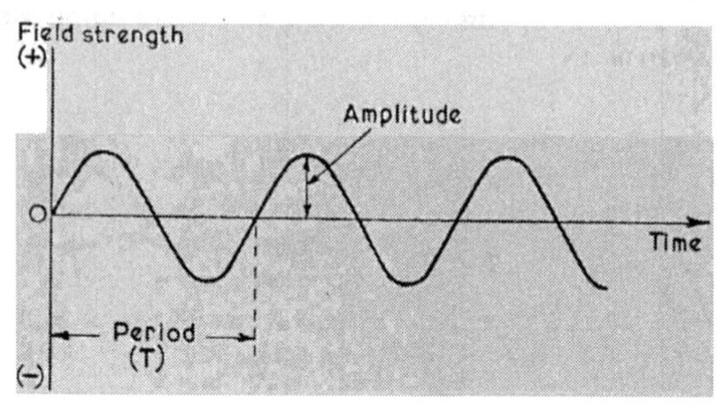

FIG. 1.13

forming the origin of electric or *Hertzian waves*. The wireless energy which is radiated from a transmitter is said to travel with wave motion. Associated with the moving wireless energy are two fields of disturbance: one an alternating magnetic field, and the other an alternating electric field. Represented graphically, as illustrated in Fig. 1.13, either of these disturbances produces a sine curve. The maximum displacement of the sine curve from the zero line of the field strength represents the *amplitude* of the waves.

Herein lies the reason for using the term *wave motion*. It should be appreciated that, when wireless energy is travelling from a transmitter, no changing shape of any surface is involved as is the case with water waves.

The distance travelled by radio energy during one oscillation of the electrons in the oscillatory circuit from which the energy is being radiated, that is, in the period or time taken for one cycle to be completed, is known as the *wavelength*. This is normally given in metres and is represented by the Greek letter lambda (λ).

If the period or frequency of the transmitted radio energy is known, the wavelength may readily be found, knowing the velocity (v) to be 300 million metres per second. Suppose f cycles are made in one second. Then during one second, f complete waves will have been transmitted. The wave made at the beginning of the one-second interval will be 300 million metres from the transmitter and the one made at the end of the one-second interval will be at the transmitter. Spread over this distance there will be f waves uniformly distributed, and the wavelength will therefore be 300 million/f metres. In other words the product of the frequency (f) in cycles per second (c/s), and the wavelength (λ) in metres is equal to 300,000,000; that is

$$f\lambda = 300,000,000$$

If the frequency is given in kilocycles per second (1 kc = 1,000 c), the product of the frequency and the wavelength will be 300,000: that is

$$f\lambda = 300,000$$

The complete range or gamut of radio frequencies form what is referred to as the *radio-frequency band.* This is divided arbitrarily as follows—

Name	*Frequency*	*Wavelength*
Very low freq. (V.L.F.)	<30 kc/s	>10,000 m
Low freq. (L.F.)	30– 300 kc/s	10,000–1,000 m
Medium freq. (M.F.)	300– 3,000 kc/s	1,000– 100 m
High freq. (H.F.)	3– 30 Mc/s	100– 10 m
Very high freq. (V.H.F.)	30– 300 Mc/s	10– 1 m
Decimetre waves	300– 3,000 Mc/s	100– 10 cm
Centimetre waves	3,000–30,000 Mc/s	10– 1 cm

The frequency of transmitted radio energy is governed by the characteristics of the oscillatory circuit which forms the essence of the transmitter. It may be determined by means of the formula

$$f = \frac{1}{2\pi\sqrt{LC}}$$

where L is the inductance of the circuit in henrys and C is the capacitance in farads.

Radio energy is radiated from an *aerial,* which in its simplest form is a vertical wire which is said to have *capacitance to earth.* This means simply that the aerial and the Earth act in much the same way as the plates of a charged condenser, and between them an electrostatic field exists when the transmitter is in operation.

Fig. 1.14 illustrates the principle of radio transmission. The oscillatory circuit shown here is of the open type, as opposed to the closed oscillatory circuit described earlier. The radiation of wireless energy from the aerial persists so long as the energy lost to the circuit is made up by that which is provided by the circuit itself.

Resonance

The frequency of any oscillatory circuit is dependent upon the capacitance and inductance of the circuit (as explained above). For any given values of capacitance and inductance, the frequency

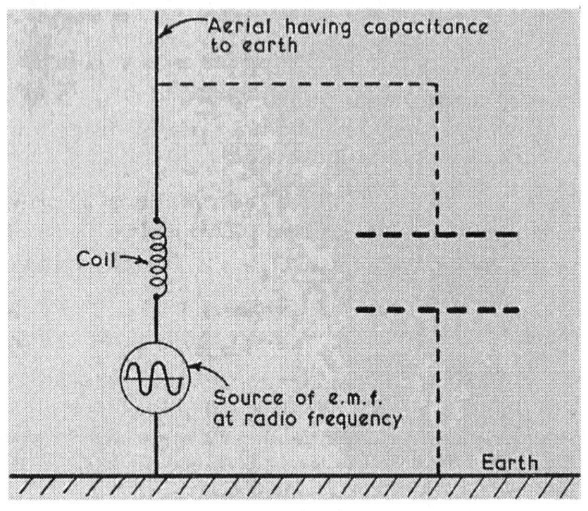

Fig. 1.14

of the circuit is referred to as its *natural frequency*. In an open oscillatory circuit which is radiating radio energy continuously, the variable current in the circuit attains greatest amplitude when the frequency of the applied e.m.f. is the same as the natural frequency of the circuit. The circuit is said to *resound* to the applied e.m.f. when the frequency of the e.m.f. is the same as the natural frequency of the circuit, just as a stretched string resounds to an impulse of suitable frequency. Another useful analogy is to be found in synchronism—a phenomenon well known to sailors.

The frequency of an oscillatory circuit may be adjusted by altering its capacitance (normally by means of a variable air condenser), or by altering its inductance (by means of a variometer). When the frequency of an oscillatory circuit has been made equal to that of the applied e.m.f., the circuit is said to be at *resonant frequency*. The process of achieving this condition is known as *tuning*.

A circuit is said to be tuned when the reactance of the condenser is equal to the reactance of the inductance of the circuit.

The reactance of the condenser is the resistance it offers to alternating current (a.c.), and is dependent upon the frequency and capacitance of the circuit.

$$\text{Reactance of condenser } (X_C) = 1/2\pi fC$$

The reactance of the inductance is the resistance the inductance of the circuit offers to a.c., and is dependent upon frequency and inductance.

$$\text{Reactance of inductance } (X_L) = 2\pi fL$$

For resonance

$$X_C = X_L$$

i.e.

$$2\pi fL = 1/2\pi fC$$

from which

$$f = 1/2\pi\sqrt{LC}$$

Radiation from Transmitting Aerial

Fig. 1.15 illustrates the electric and magnetic fields which develop around a transmitting aerial. The aerial carries alternating current of radio frequency, and having capacitance to Earth it becomes alternately positively and negatively charged. At the instant when the aerial is positively charged, a field of strain exists around it in the form of lines of electrostatic force. The electrolines repel one another but are prevented from expanding by their tendency to reach the Earth by the shortest possible route. Fig. 1.15(*a*) represents a vertical section through the aerial when the aerial is positively charged. An instant later when the current is down the aerial, the electrolines tend to collapse on to the aerial and the energy associated with them returns in consequence to the circuit. As the electrostatic field decays, a magnetic field develops around the aerial, and reaches maximum intensity, as illustrated in Fig. 1.15(*b*), when the electrostatic field has zero intensity. After this instant the aerial becomes negatively charged and an electrostatic field again develops but this time in a direction opposite to that of the electrostatic field formed when the aerial was positively charged. As this field develops the magnetic field decays, and the energy associated with the lines of magnetic force tends to return to the circuit. The electrostatic field reaches maximum intensity at the instant when the magnetic field ceases momentarily to exist. An instant after this state occurs, the current in the aerial has a direction up the aerial and in consequence a magnetic field develops, the direction of which is

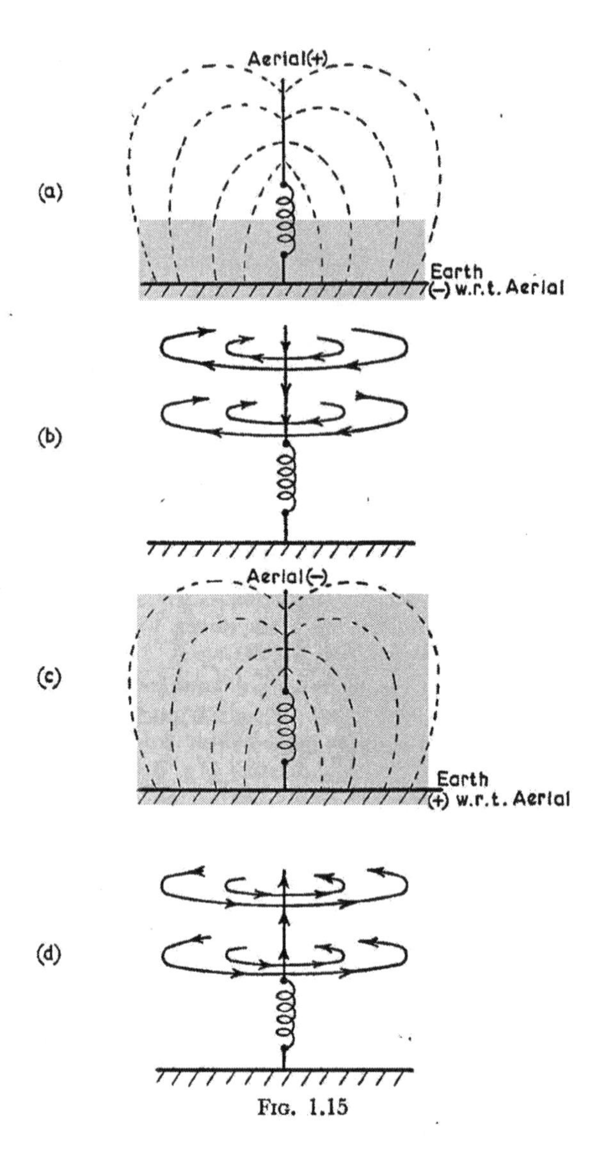

Fig. 1.15

opposite to that formed when the current was down the aerial. This magnetic field reaches maximum intensity at the instant the electrostatic field becomes zero in intensity. This condition is illustrated in Fig. 1.15(*d*). As the aerial again becomes positively charged the magnetic field decays, and an electrostatic field grows until the condition of the aerial is the same as what it was initially, as depicted in Fig. 1.15(*a*). At the instant when this occurs, one complete cycle of the alternating current or p.d. between aerial and Earth is said to have been made. The current and p.d. are, it will

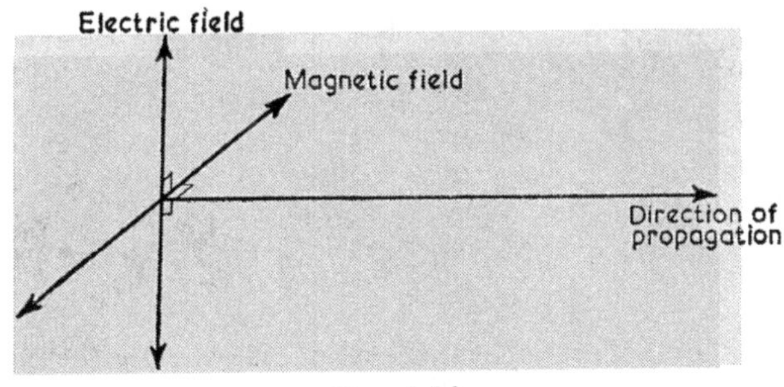

FIG. 1.16

be noticed, a quarter of a cycle, or 90 degrees, out of phase with one another. Fig. 1.15, together with the remarks pertaining to it, should be compared with the notes relating to the closed oscillatory circuit illustrated in Fig. 1.11.

If the frequency of the oscillations in the type of circuit described is sufficiently great, that is to say, if the frequency is within the radio-frequency waveband, a certain proportion of the energy associated with both the electrostatic and magnetic fields is prevented from returning to the aerial circuit, as each field decays. The lines of force associated with this "lost" energy are therefore crowded outwards in all directions from the aerial, this resulting in the radiation or propagation of electromagnetic energy.

The radiated energy is associated with a combination of electric and magnetic fields, the directions of which are at right-angles to one another, and also at right-angles to the direction of propagation, as illustrated in Fig. 1.16.

When the electrostatic field which is associated with radio energy is vertical, the propagated waves are said to be *vertically polarized.* When the electrostatic field is horizontal the waves are *horizontally polarized.* The *plane of polarization* is the plane which contains both the electrostatic field and the direction of propagation.

The strengths of the alternating fields at varying distances from the transmitter may be represented graphically as in Fig. 1.17.

Within a short distance of a transmitter using a vertical aerial, both the electric and magnetic fields are appreciably curved, but as the distance from the transmitting aerial increases the curvature becomes slighter, such that the network of lines of force near the Earth's surface lies in almost a vertical plane, presenting a vertical *wavefront.*

The amount of energy propagated in a unit of time is dependent upon the *power* of the transmitter, and this affects the frequency and

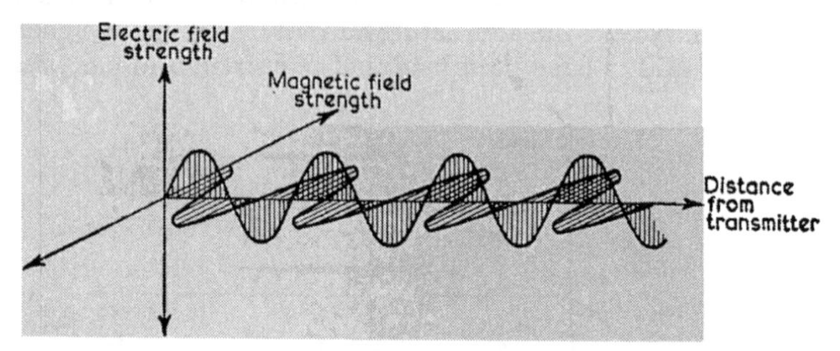

FIG. 1.17

amplitude of the propagated waves. The greater the power of the transmitter, the greater is the amplitude and/or the frequency of the radiated energy.

The range at which radio energy may be detected by a suitable receiver is dependent upon the power of the transmitter and also upon the height of the transmitting aerial.

Propagation of Radio Waves

All radio energy tends to travel along the Earth's surface, but V.L.F. energy has a greater tendency to do so than energy of higher frequency. V.H.F. and centimetre and decimetre waves tend to maintain the directions in which they are travelling when they leave the transmitting aerial. Those waves that travel initially in directions other than the horizontal are not therefore drawn towards the Earth's surface to such a great extent as waves of lower frequency. For this reason the range at which centimetre waves (such as are used in radar) are detected is only slightly more than optical range.

The wavelength used by Hertz in his original experiments with radio energy was about three metres. The scientists who developed Hertz's discovery—foremost amongst whom was Marconi—found that longer wavelengths gave greater ranges, and from about 1895 onwards it was considered that long waves, that is, low-frequency transmissions, were more suitable for long-distance communication than short-wave or high-frequency radio energy.

After the Great War, increased attention was paid to short wavelengths. The invention of the thermionic valve by Fleming in 1904, and its development by Lee de Forest in 1907, provided new and powerful tools in the hands of those engaged in research into the transmission and reception of radio energy.

Following considerable experimentation with valve transmitters and receivers, it was found that wireless energy of relatively short wavelength travelled considerable distances with relatively small loss of strength, or, as it is technically described, with little *attenuation*. It was also discovered that the field strength of radio energy of short wavelength decreased rapidly as the distance from the transmitter increased, but suddenly increased to relatively great strength when a certain critical distance, now known as the *skipped distance*, was reached.

The skipped distance was found to vary with the wavelength of the transmitted energy. For example, using a wavelength of 30 m it was found to be about 600 km, and using 20 m wavelength it was found to be 1,400 km.

Although, as stated earlier, radio energy tends to travel parallel to the Earth's surface, radiation from a vertical aerial is emitted in all directions, not only in the horizontal plane. The proportion of radiated energy does, however, decrease with increasing angle of elevation above the horizon, such that most of the radiated energy is within about 30 degrees of the horizontal plane at the transmitter. The energy which travels adjacent to the Earth's surface diminishes in strength, that is it becomes attenuated, as the distance from the transmitter increases. If the Earth had no electrical resistance, that is, if it were a perfect conductor, the induced charges, due to the electric and magnetic fields associated with the radio energy, would travel along with the energy and no attenuation would result. The Earth's resistance depends upon the physical nature of the surface, being less for water than for land. As the attenuation varies inversely as the resistance (or directly as the conductance), the range over the sea is greater than it is over the land for any given transmitter.

The radio energy which moves away from the transmitter aerial at an angle to the horizontal is sometimes returned to the Earth's surface after being reflected from an electrically charged region in the Earth's atmosphere, which acts as a mirror to radio waves. This region is known as the *ionosphere*, and within it are to be found regions known as *ionized layers*. Within the ionized layers there is an abundance of free electrons due to bombardment of the gases in the atmosphere at certain heights above the Earth's surface, through the agency of extremely high-frequency electromagnetic radiation emanating from the stars and space. Such radiation is known as *cosmic radiation*, and the greater proportion of the total cosmic radiation reaching the Earth's atmosphere originates in the Sun—the nearest star to the Earth.

Ionized layers have two very important effects on radio energy. Firstly they cause refraction and possible reflexion of radio energy, and secondly they absorb a fraction of any radio energy which passes through them, thus causing attenuation.

The degree of ionization within the ionized layers varies with time. The time of day, time of year, and time in the eleven-year cycle of sunspots all are significant factors, the Sun being the principal agency by means of which ionization of the atmosphere occurs.

When the Sun is above the horizon of any part of the Earth's surface, ionization is intense, and a considerable proportion of energy radiated skywards from a transmitter is absorbed, only a small amount being reflected back to the Earth. When the Sun is below the horizon, the degree of ionization is slight, absorption of energy is therefore small, and a large proportion of the total energy is reflected back to the Earth.

As well as a day and night variation in the degree of ionization, there is a seasonal variation. During the winter in high latitudes, especially when the Sun has a low altitude, ionization is slight, and strong reflexions from the ionized layers may result during daytime as well as during night time.

Signals received at a receiver, due to radio energy which has travelled over the Earth's surface between the transmitter and the receiver, are referred to as *ground-wave signals*. Those due to reflected energy are known as *sky-wave signals*.

As the range from a transmitter increases, ground-wave signals become weaker because of attenuation. The maximum distance at which ground energy may be received is known as the *ground range*. Except at short ranges, sky-wave signals during the night time are usually stronger than ground-wave signals, since, although the reflected energy has to travel a greater distance than the ground energy, it is considerably less attenuated. The distance between the transmitter and the nearest point to it at which sky waves are received is the skipped distance, referred to earlier.

Fig. 1.18 illustrates the refraction and reflexion of radio energy at the ionosphere. It will be noticed that the refraction increases as the angle which the incident energy makes with the vertical increases. When this angle is greater than a critical angle (θ in the figure), at which the refracted energy is parallel to the horizon, reflexion at the ionosphere takes place.

When the range is such that it is impossible for a receiver to receive sky-wave radiation, the receiver is within the skipped distance, and is said to be *in the skip*.

Should the ground range be less than the skipped distance, there

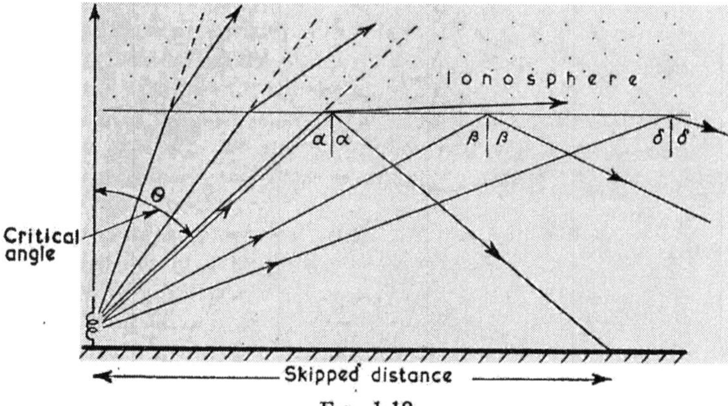

FIG. 1.18

will be a ring-shaped zone centred at the transmitter, in which no radio energy from the transmitter may be received. This region is known as *dead space*, and is illustrated in Fig. 1.19.

It will be noticed from Figs. 1.18 and 1.19, that sky waves are, in general, not vertically or horizontally polarized.

Following the researches into the propagation of radio energy, in the early part of the century, two scientists, the American Kenelly and the Britisher Heaviside, discovered independently of one another, in 1902, that radio energy may be reflected from what they believed to be an ionized region high above the Earth's surface. Now it is well known that gases when ionized are good electrical conductors, whereas at normal temperature and pressure they are excellent insulators. This conductive property of ionized gas led Kenelly and Heaviside to predict that the reflecting layer was one in which certain constituent gases of the atmosphere were highly ionized. Although the word layer is misleading, since the reflecting region has no distinct boundary surfaces, the first reflecting region discovered is now known as the Kenelly-Heaviside layer, in honour of the two scientists who discovered its presence. This ionized layer has a variable height of between roughly 50 and 100 miles above the Earth's surface. After sunset, at which time there is a sudden reduction in cosmic radiation, the free electrons in the lower levels of the ionized layer recombine with unstable atoms, with the result that the ionized layer rises, to fall again after sunrise on the following day.

Another reflecting region was discovered in 1925. This is the

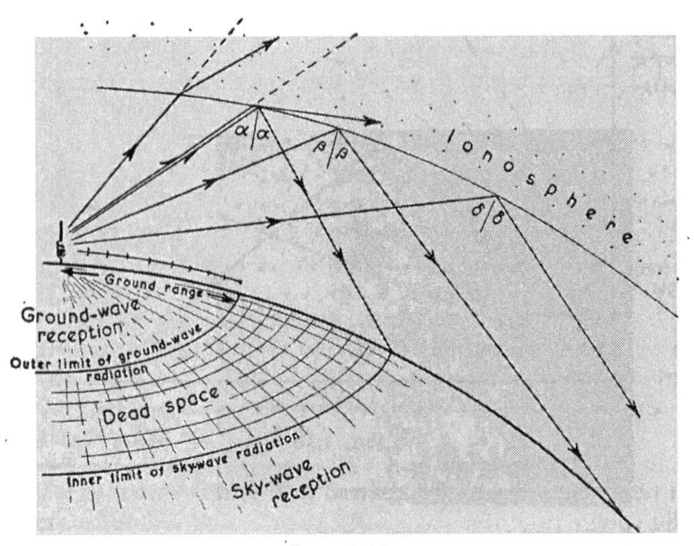

Fig. 1.19

Appleton layer, named after its discoverer. Its height varies between 100 and 250 miles above the Earth's surface.

Reception of Radio Waves

Wireless energy is detected or *picked up* by a device known as a *radio receiver*. A receiver is similar in many respects to a radio transmitter. Again, as in the transmitter, the essential features of a receiver are an aerial and an oscillatory circuit. The receiver is first *tuned* to the frequency on which the transmitter is working. This, it will be remembered, is done by adjusting the capacitance or inductance of the receiver circuit for resonance. The electromagnetic energy, on reaching the receiving aerial, induces an alternating current—in accordance with Faraday's law of induction—in the aerial circuit. Since the density of the lines of force (known as the *flux density*) associated with the radio energy changes as the radio energy moves past the receiver aerial, the e.m.f., which causes the induced current, has the same frequency as that of the changing flux, which is the same as the frequency of the transmitter circuit. The greater the rate of change of flux density, the higher is the frequency and the greater is the amplitude, and hence the greater will be the induced e.m.f. The longer the receiver aerial, the greater will be the flux density—or number of lines of force cutting it—and, accordingly, the greater will be the induced e.m.f.

The alternating current which is induced in the receiver aerial is used to create a signal—usually audible—the strength or loudness of which is dependent upon the sensitivity of the receiver.

An audible signal is produced by a process known as *heterodyning*, a process in which the alternating current induced in the receiver aerial is combined or *mixed* with an alternating current provided by means of an oscillatory circuit—known as a *local oscillator*—the characteristics of which are adjusted so that it produces an alternating current at a frequency slightly different from that of the current induced in the aerial circuit.

The result of combining two alternating quantities of different frequencies is to produce what is known as a *beat frequency*, equal to the difference between the frequencies of the alternating quantities that have been mixed. A simple example of heterodyning, familiar to navigators, is the combining of the solar and lunar tides. The solar tide has a period of 12 hours, and its frequency therefore is 60 per 30 days. The lunar tide has a period of approximately 12½ hours, and its period therefore is, in round figures, 58 per 30 days. The difference between these two frequencies is 2 per 30 days, and this, of course, is the frequency of *springs* or *neaps*.

The alternating current at beat frequency energizes the coil which is incorporated in a telephone or loudspeaker, and the resulting alternating magnetic field causes the diaphragm of the instrument to oscillate; this in turn causes the air in the vicinity to vibrate with the result that sound energy is produced, the pitch of the note being determined by the beat frequency, which is made to fall within the audio-frequency band.

The human ear is automatically tuned to detect or receive sound energy, just as a wireless receiver may be tuned to receive radio energy of a specified radio frequency. The approximate limits of audio frequency are 30 and 15,000 cycles per second. The frequency of middle C, for example, is 256 cycles per second, and that of the note an octave higher is 512 cycles per second.

The term *modulation*, applied to radio, means an alteration in the frequency or amplitude of a radio carrier wave. The simplest type of radio transmitter produces an alternating current of constant radio frequency and constant amplitude. A receiver tuned to such a transmitter would produce, after heterodyning, a constant and uniform sound signal, the frequency and loudness of which would be dependent upon the beat frequency and amplitude, after

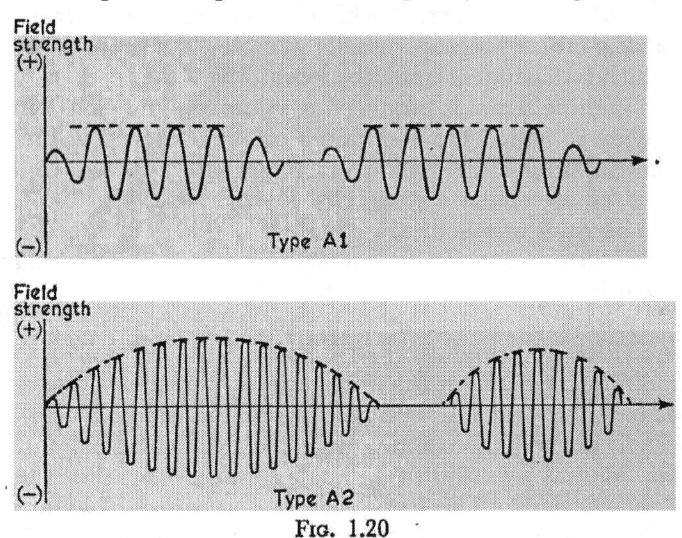

Fig. 1.20

mixing the incoming waves with those produced by the local oscillator.

In order to convey messages by radio means, the alternating radio-frequency current in the transmitter aerial is interrupted by a process known as *keying*, by means of which messages are transmitted, utilizing the morse code to do so. Such transmitters are said to produce continuous waves (C.W.) of a type known as A1. Waves of the type A1 necessitate the use of a heterodyning device at the receiver, before signals can be rendered perceptible to the ear. A1 waves are described as: *continuous waves; unmodulated; key controlled.*

An alternative method of conveying messages by radio in which heterodyning at the receiver is not necessary, is to modulate the radio-frequency waves at the transmitter, by a variation in amplitude, keying providing for the transmission of messages. Transmitters working on this system are said to be using waves of the type A2. A2 waves are described as: *continuous waves; amplitude modulated at audible frequency; key controlled.*

Fig. 1.20 illustrates graphically waves of types A1 and A2.

Part 2

The Principles of Radio Direction Finding

THE radio direction finder ranks as one of the earliest radio aids to navigation, standing second in chronological order to the radio time signal.

The first experiments on wireless direction finding were conducted in 1908, and in 1911 the first ship—the Mauretania—was fitted with wireless direction finding gear. From about 1920, following the introduction and development of thermionic valves, rapid progress in the advancement of radio techniques, and accordingly in the science of direction finding by wireless, was made. In 1935 certain ships were required by law to be equipped with wireless direction finders, and in 1948 this requirement was extended, so that at the present time all sea-going ships of 1,600 G.R.T. and above, and all passenger ships regardless of their tonnage, are to be fitted with an efficient radio direction finder.

A radio direction finding system may be regarded as being comprised of three distinct units. These are—

1. A shore-based transmitting station.
2. A handbook from which the service details of the transmitting stations may be obtained.
3. A ship-based direction finding instrument.

There are two main types of transmitting station used in the practice of radio direction finding. These are known as *Radio Direction Finding Stations* and *Radio Beacons*.

A radio D.F. station is a shore-based radio station which is equipped with wireless direction finding gear. Such stations are therefore capable of determining radio bearings of ships at sea. There are relatively few radio D.F. stations compared with the large number of radio beacons. Radio D.F. stations provide services enabling ships—especially those which are not equipped with radio direction finding gear—to ascertain their positions when the normal visual or astronomical means are not available, or when they are not suitable.

Some radio D.F. stations are equipped with transmitting and

receiving gear, and operate quite independently. In other cases, several radio D.F. stations work as a team. Each member of such a group is equipped with a direction finder, but only one—with whom the remainder are linked by telephone—is provided with transmitting gear.

A radio beacon is a radio station, located on shore, at a lighthouse, or on a lightship, which is equipped with transmitting gear. Signals are transmitted, by means of which ships at sea are able to determine the bearing of the transmitter by using their direction finders. Although radio beacons are expressly designed as aids to navigation when visibility is poor, and the normal visual and astronomical methods of navigation are then not possible, many beacons do operate in clear weather as well as in conditions of poor visibility.

Under the terms of an international telecommunications convention which met in 1947, authorities which establish a direction finding service are expected to ensure that the service is effective and regular. Mariners are warned, however, that no responsibility will be shouldered by an authority in respect of any consequences that might arise from the use of any information they may provide, or for the defective working or failure of any of their stations.

Service details and general instructions and information, relating to wireless direction finding stations and radio beacons, are to be found in *The Admiralty List of Radio Signals*, Volume 2—the handbook which forms an important part of a radio direction system.

From the navigator's viewpoint the ship-based radio direction finder is perhaps the most important part of a radio direction system.

The Loop Aerial

Direction finding by means of radio is possible through the agency of a particular form of receiving aerial known as a *loop aerial*. In its simplest form a loop aerial is nothing more than a coil of wire set up so that its plane is vertical, the coil being capable of being rotated about its vertical axis.

Consider the vertical coil of wire mounted on a ship as illustrated in Fig. 2.1 (*a*) and (*b*). The horizontal magnetic field created by a distant transmitter will, on passing the coil, set up in it, in accordance with Faraday's law of magnetic induction, an induced e.m.f. As a result of this induced e.m.f. an induced alternating current will develop in the loop circuit, provided that the alternating magnetic field of the transmitted radio energy threads the loop. This will always be the case, so long as the plane of the loop is not at right-angles to the direction of the transmitting station. Should the plane of the loop lie at 90° to the direction of the transmitter, the changing magnetic field will be coplanar with the loop and there will be zero e.m.f. induced in the loop. This condition is illustrated in Fig. 2.1(*a*). If the plane of the loop lies in the direction of the transmitting station, as illustrated in Fig. 2.1(*b*), the induced e.m.f. in the loop will have maximum value. If the loop is rotated from the position in which the induced e.m.f. is maximum, through a

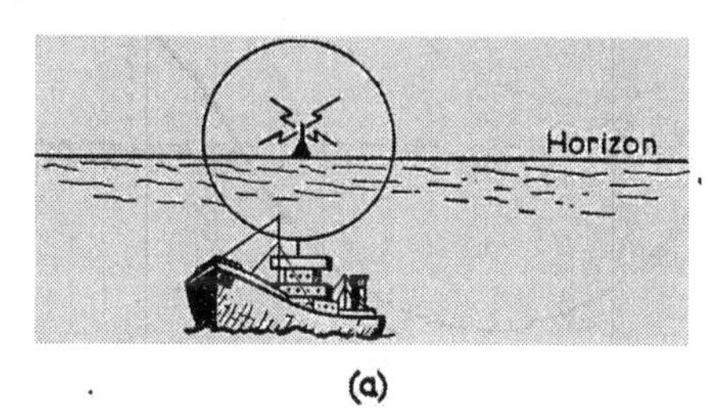

(a)

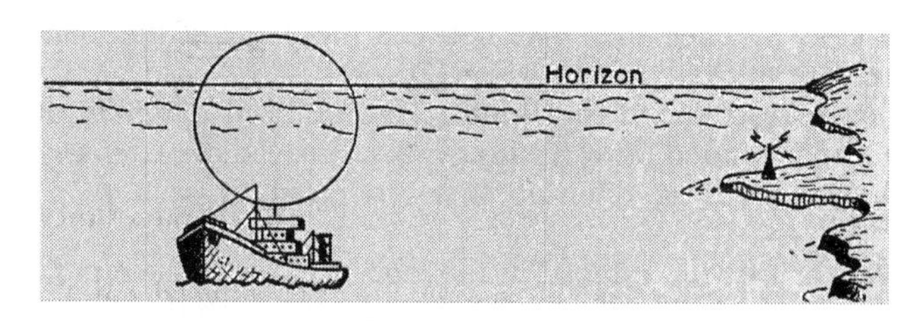

(b)

Fig. 2.1

complete circle, the induced e.m.f. will change, its value at any position being dependent upon the cosine of the angle between the plane of the loop and the direction of the transmitting station.

Fig. 2.2 represents graphically the e.m.f. induced in the loop, against the angle between the plane of the loop and the direction of the transmitter (Tx). It will be noticed that the induced e.m.f. is zero at two positions during the rotation, viz. when the angle between the plane of the loop and the direction of the station is 90° or 270°; and that the induced e.m.f. has maximum value at two positions during the rotation, viz. when the angle between the plane of the loop and the direction of the transmitting station is 0° or 180°.

When the angle between the plane of the loop and the direction of the station is 180°, for any given changing field threading the loop, the direction of the induced current in the loop is the reverse from its direction when the angle between the plane of the loop and the direction of the station is 0°.

At any angle θ, between 90° and 270° through 180°, the induced current in the loop is said to be 180° out of phase with the induced

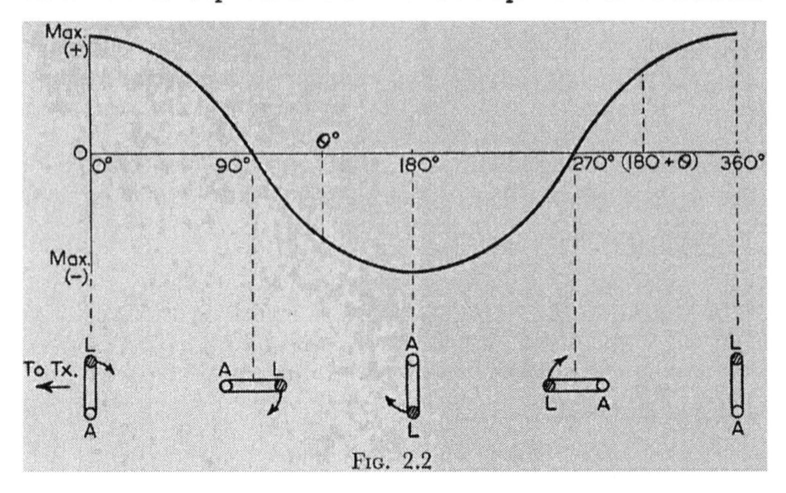

FIG. 2.2

current in the loop, when the angle between the plane of the loop and the direction of the transmitter is $(180 + \theta)$.

An alternative method of representing the induced e.m.f. in a loop, as the loop is rotated, is by means of polar, instead of cartesian, coordinates, the resulting diagram being referred to as a *polar diagram.*

If the maximum e.m.f. induced in the loop is E_m, and the e.m.f. induced in the loop when its plane makes an angle θ with the direction of the transmitter is E_θ, then

$$E_\theta = E_m \cos \theta$$

The polar diagram, showing the magnitude of E_θ for all values of θ from 0° to 360°, is illustrated in Fig. 2.3. The value of E_θ for any given value of θ is plotted as follows. Referring to Fig. 2.3, the arm OA is rotated anticlockwise through the angle θ, and the value of E_θ is marked off along this arm from O. If this is done for all values of θ from 0° to 360°, the polar diagram, as illustrated, is in the form of two circles, and is usually referred to as a *figure-of-eight diagram.*

The polar diagram of the curve $E_\theta = E_m \cos \theta$ is, in fact, a single circle to the right of the line XY in Fig. 2.3. The figure-of-eight diagram, however, provides a convenient method of showing not only the magnitude of the e.m.f. induced in the loop, but also the phase of the induced e.m.f. The figure-of-eight diagram is, strictly, the polar diagram of the equation

$$E_\theta = \sqrt{(E_m \cos \theta)^2}$$

The polar diagram representing the e.m.f. induced in a vertical aerial is a circle centred at the origin O, indicating uniform recep-

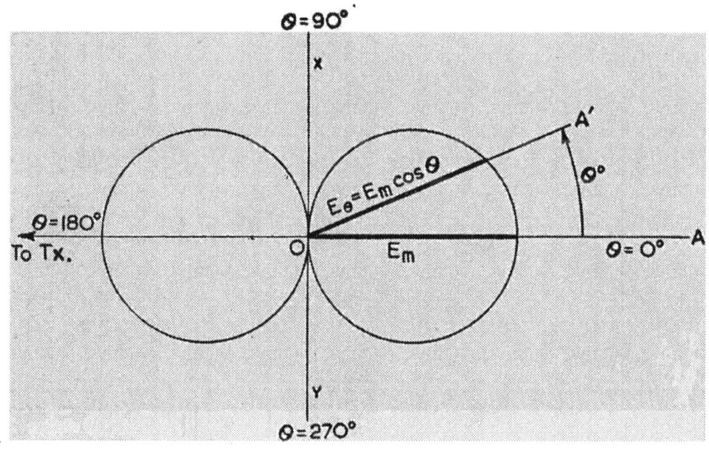

FIG. 2.3

tivity from all directions. The induced e.m.f. in a vertical aerial is in phase with the changing field causing the induction, whereas the induced e.m.f. in a loop aerial is 90° out of phase with the field. This means that, when the field has maximum intensity, its rate of change of intensity is zero, and, as will be remembered, the induced e.m.f. is a function of the rate of change of the strength of the threading field.

The greater the number of turns in the loop aerial and the greater the area enclosed by the loop, the greater will be the e.m.f. induced in the loop aerial. This follows from the fact that the induced e.m.f. in a coil is proportional to the rate of change of the magnetic field threading the coil and the magnitude of the flux density.

In a radio direction finder of the simplest type, the loop aerial is capable of being rotated about a vertical axis. The current induced in the loop aerial is used to produce an audible signal, the receiver circuit first being tuned to the frequency of the transmitter, and the sound signal created by heterodyning if necessary.

As the loop aerial is rotated, maximum signal strength is obtained when the plane of the loop aerial lies in the direction of the transmitting station, the signal strength falling off as the loop aerial is rotated away from this direction. A graph of the signal strength against the angle between the plane of the loop and the direction of the transmitter is illustrated in Fig. 2.4.

Examination of the curve illustrated in Fig. 2.4 reveals that the rate of change of signal strength is least when the signal strength is

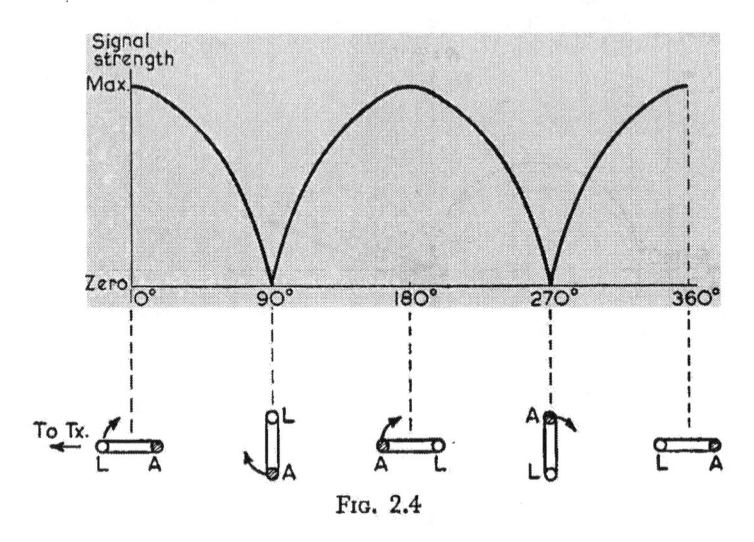

FIG. 2.4

greatest. This occurs when the angle between the plane of the loop aerial and the direction of the transmitter is 0° or 180°. The rate of change of signal strength is greatest when the signal strength is least, that is, at the positions when the loop plane is at 90° or 270° to the direction of the transmitter. In other words, as the aerial is rotated, the minimum signal strength position is approached very much more sharply than is the position for maximum signal strength. Near the position for maximum signal strength, a large change in the direction of the plane of the loop aerial results in a relatively small change in the signal strength, whereas, near the position for minimum signal strength, a small change in the direction of the plane of the loop aerial results in a relatively big change in the signal strength. For this reason it is easier to determine the position of the loop aerial for the minimum or zero signal strength than it is to determine the position of the loop aerial for maximum signal strength.

The angle between the plane of the loop aerial and the direction of the transmitting station is indicated by a *bearing pointer*, which moves, as the aerial is rotated, within a graduated compass card. Because it is easier to discriminate the position of the loop for minimum, than it is for maximum, signal strength, the bearing pointer is placed at right-angles to the plane of the rotating loop aerial.

In a direction finder which has a rotating loop aerial, the compass card is set such that the 0° and 180° directions represent right ahead and right astern respectively; that is, when using the instrument, the bearing pointer will indicate the *relative bearing* of the transmitter.

In order to determine the bearing of a radio beacon or other radio transmitter, by means of a rotating loop direction finder, the receiver is switched on, and the headphones are plugged in. The receiver is then tuned to the frequency of the transmitter, the bearing of which is required (this being determined from *The Admiralty List of Radio Signals*, Volume 2, or a similar publication). The loop aerial is then rotated as the volume or *gain* control is adjusted until the signal becomes audible, and the position of the loop is set for a zero signal. This applies when the transmitter is working with waves of the A2 type: if A1 type waves are being transmitted it will be necessary, before an audible signal can be produced, for the unmodulated waves that are being received to be heterodyned.

Setting the loop for the zero-signal position is normally done by *bracketing the zero*, that is to say, by noting the directions indicated by the bearing pointer at two positions near the position for zero on each side of it, where the signal strengths are the same. For example, if the signal strength when the bearing pointer indicated 46° was the same as the signal strength when the pointer indicated 50°, the position for zero would be midway between 46° and 50°, that is at 48°. In practice, owing to the possibility of receiving radio energy, from transmitters other than the one to which the D.F. is tuned, of a frequency equal to or near to that of the transmitting station; or the possibility of receiving so-called atmospherics; or the possibility of *noise*, due to imperfections of the receiver itself, it may happen that when the pointer indicates the relative bearing of the transmitting station there is in fact an audible signal. In this event, a search is made for the position of the bearing pointer when a minimum signal is being received.

Generally, when obtaining a wireless bearing of a radio beacon or other transmitting ship or shore station, the approximate bearing of the station is known, and accordingly, when determining its bearing, the bearing pointer is swung to and fro in the region of the approximate relative bearing, and a search is made for the zero-signal position in this region. It will be remembered that there are two directions—separated by 180°—in which the plane of the loop aerial may be set so that a zero or minimum signal results. Occasionally when using a radio direction finder to ascertain a radio bearing, the approximate direction of the transmitting station is not known. This may be the case, for example, when it may be necessary to use the direction finding apparatus for the purpose of *homing* on a vessel in distress, the position of which is near to a potential rescue ship. When the approximate direction of the transmitter is not known, it will be necessary to resolve the 180° *ambiguity* which arises from there being two positions of the loop aerial for a zero or minimum signal.

The Sensefinder

The device provided to resolve the 180° ambiguity is known as a *sensefinder*, and the process of doing so is known as *sensing*. A simple direction finder, which is not fitted with a sensefinder, may be used to determine the vertical plane of the direction of the radio energy as it passes through the loop aerial, but such a direction finder cannot be used to determine the *sense* of the signal, that is to say, the direction from which the energy has come is indeterminate by the instrument alone.

The sensefinder is fitted with a vertical aerial which is separately connected to the amplifying unit of the D.F. receiver. Now the e.m.f. induced in a vertical aerial is not affected, either in amplitude or phase, by the direction of the transmitter to which it is tuned. The loop aerial e.m.f., however, is affected, the e.m.f. induced in a loop aerial being equal in magnitude but opposite in phase to the e.m.f. induced by a similar transmitter on the reciprocal bearing.

When it is necessary to sense, the sensefinder is switched on, and the signal heard is then due to a combination of the e.m.f.s induced in the loop and vertical aerials.

The characteristics of the sensefinder circuit are adjusted so that the e.m.f. induced in the vertical aerial is equal in magnitude to the maximum e.m.f. induced in the loop aerial. Whereas the curve of e.m.f. induced in the loop aerial against the angle which the plane of the loop makes with the direction of the transmitting station is a cosine curve, the curve of e.m.f. induced in a vertical aerial is a straight line, as the direction of the transmitter does not affect the magnitude of the e.m.f. induced in it.

The graphs of the loop and sense aerial e.m.f.s are depicted in

Fig. 2.5. The resultant e.m.f.—obtained when the sensefinder is switched on—is also shown. The resultant e.m.f. is derived by combining the e.m.f.s induced in the loop and vertical aerials.

When the sensefinder is switched on, and the loop aerial is rotated through 360°, the signal strength varies such that there is a

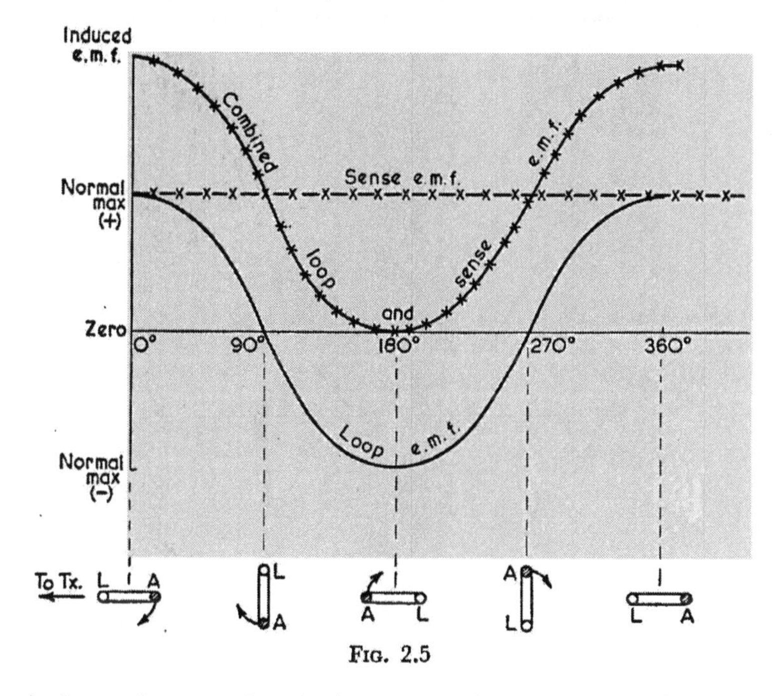

FIG. 2.5

single maximum and a single zero or minimum signal, the single maximum occurring 90° away from the orientations of the loop aerial for minimum or zero signal, using the loop aerial e.m.f. only. It is thus possible to discriminate between the zero position for the correct bearing and the other zero position. Fig. 2.6 illustrates the curve of signal strength due to a combination of the e.m.f.s induced in the loop and sense aerials, for all positions of the loop aerial.

Fitted to the spindle which carries the bearing pointer is a second pointer which is fitted 90° away from the position of the bearing pointer. This is the sense pointer and is used only when sensing. The following example serves to illustrate the principle and operation of sensing.

Suppose a radio beacon bears 50° on a vessel's starboard bow. Using a direction finder, the bearing pointer—assuming no errors in the bearing—will indicate a zero position at 50° and another zero position on the reciprocal bearing, that is, at 230°. Now if the sense pointer is pointed to 50° and 230° in turn, with the sensefinder switched on, a very loud signal will be heard for one of these positions of the sense pointer, and a weak or zero signal will be obtained on the other position. At the position where the loud signal is heard,

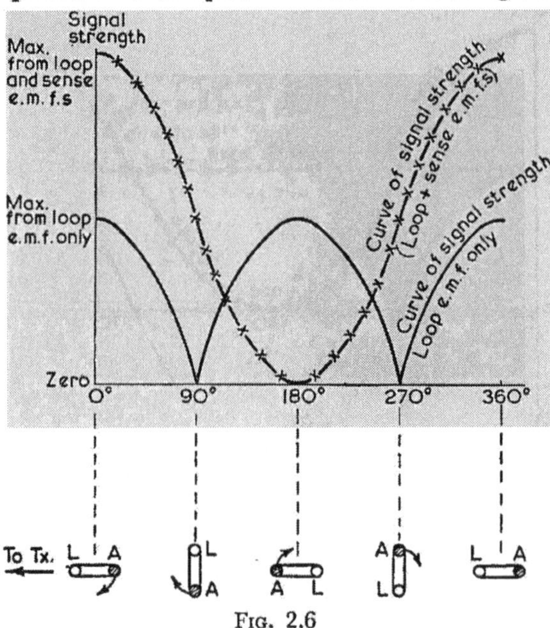

Fig. 2.6

the signal is due to the sum of the sense and loop aerial e.m.f.s, these being equal in magnitude and of the same phase as each other. At the position of the sense pointer where the weak or zero signal is obtained, the two e.m.f.s which are combined, although equal in magnitude to each other, are opposite in phase, and the resultant e.m.f. is therefore zero.

The sense pointer is fitted relative to the bearing pointer, such that the sense pointer indicates the bearing of a transmitter when a minimum signal is obtained, with the sensefinder switched on.

Fig. 2.7 illustrates this example and indicates that the bearing of the transmitting station is 50° and not 230°.

An alternative method of showing graphically the effect of combining the e.m.f.s induced in the loop and sense aerials is to employ polar coordinates instead of cartesian coordinates as has been done in Figs. 2.5, 2.6, and 2.7.

As explained earlier, the polar diagram of the e.m.f. induced in the loop aerial against the angle between the plane of the loop aerial and the direction of the transmitting station is a figure-of-eight diagram lying symmetrically to the right and left of the origin; and the polar diagram of the e.m.f. induced in a vertical aerial is

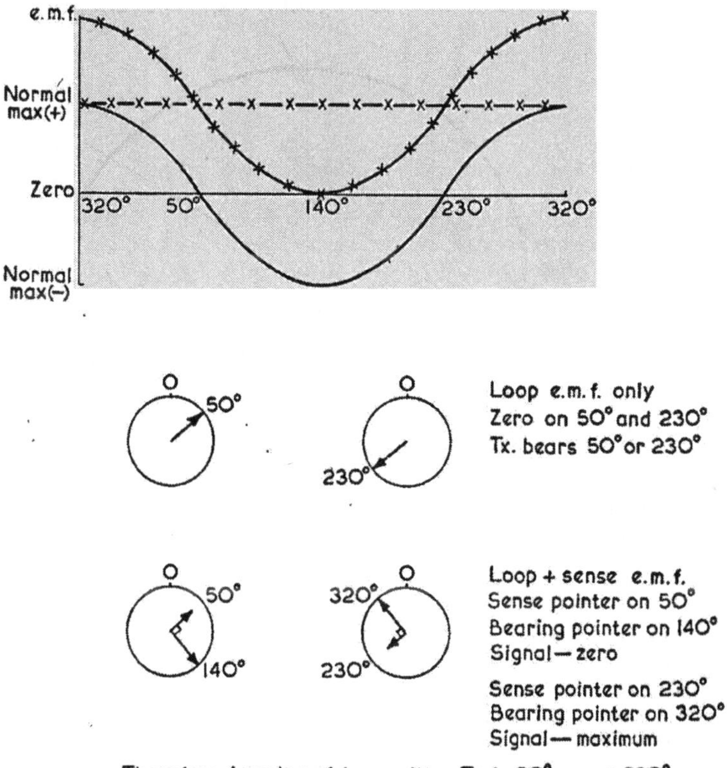

FIG. 2.7

a circle centred at the origin. If the e.m.f. induced in the vertical aerial is equal in magnitude to the maximum e.m.f. induced in the loop aerial, the figure-of-eight diagram is just enclosed in the circle representing the e.m.f. induced in the vertical aerial. The combination of the two e.m.f.s produces a heart-shaped diagram as shown in Fig. 2.8. The e.m.f.s represented in one of the circles of the figure-of-eight diagram are in phase with the e.m.f. induced in the vertical aerial, and those represented in the other circle of the figure-of-eight diagram are 180° out of phase with the e.m.f. induced in the vertical aerial. When combining the e.m.f.s, values must therefore be added on one side of the line XY (where the e.m.f.s of the loop and vertical aerials are regarded as being in phase), and subtracted one from the other on the other side of the line XY (where the two e.m.f.s are regarded as being 180° out of phase with one another).

The heart-shaped diagram illustrated in Fig. 2.8 is often referred

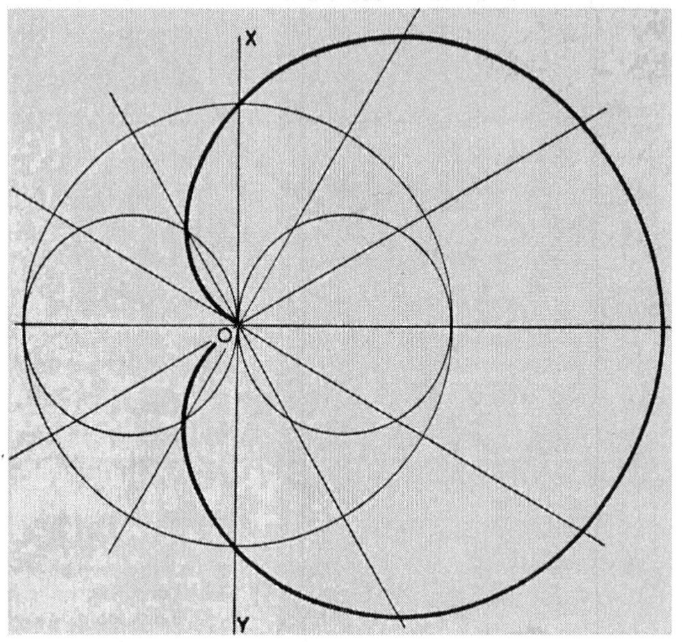

FIG. 2.8

to as a *cardioid*. It will be noticed that there is only one orientation of the loop aerial where a zero signal is obtained.

If the e.m.f. induced in the vertical aerial of the sensefinder is exactly equal to the maximum e.m.f. induced in the loop aerial, the sense pointer may be used (with the sense switch closed of course) to determine the direction of the transmitter. It is not considered good practice to do this, however, and for this reason the sense pointer does not extend to the compass graduations as does the bearing pointer. Referring to Fig. 2.7, it will be noticed that the zero signal when using the combined loop and sense aerial e.m.f.s is not approached sharply, as it is when using the loop aerial e.m.f. only. It is not therefore easy to determine the exact zero position using the sense pointer only. The chief disadvantage of using the sense pointer only, to determine the direction of a transmitter, is because of the comparative difficulty of adjusting the characteristics of the sensefinder circuit so that the e.m.f. induced in the

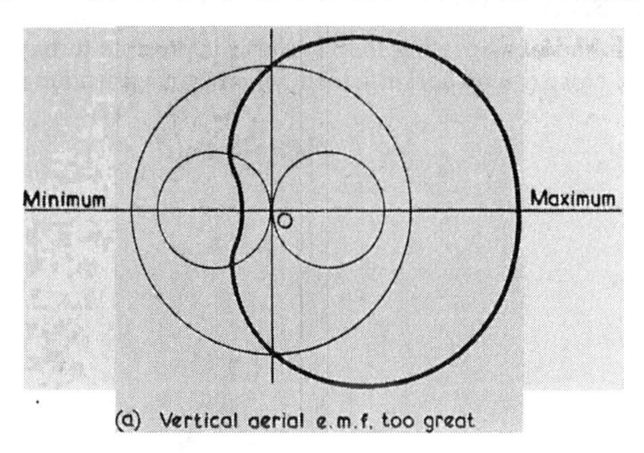

(a) Vertical aerial e.m.f. too great

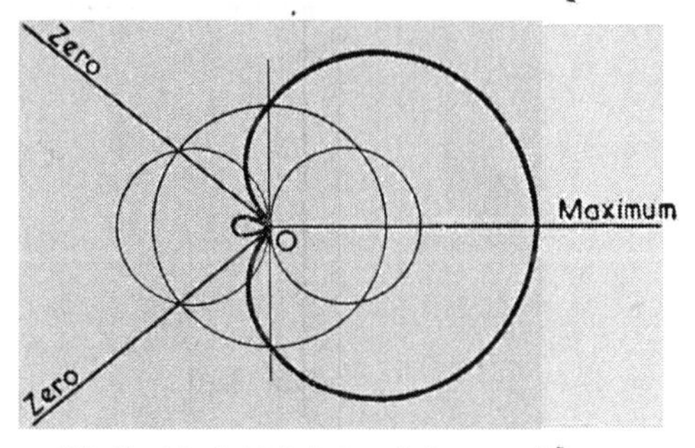

(b) Vertical aerial e.m.f. too small

FIG. 2.9

vertical aerial is exactly equal to the maximum e.m.f. induced in the loop aerial.

The polar diagrams resulting for the cases when the vertical aerial e.m.f. is less than, and greater than, the maximum e.m.f. induced in the loop aerial are depicted in Fig. 2.9. Referring to this figure, it will be noticed that, when the vertical aerial e.m.f. is too small, two zero positions are obtained, each position making equal angles with the position of the true direction of the transmitting station. When the vertical aerial e.m.f. is too large, no zero position is obtained, but a minimum signal is obtained when the sense pointer indicates the correct bearing of the transmitter, when the sense-finder is switched on.

The B.T.M. Aerial

Instead of a rotating loop aerial it is possible to have a system of fixed crossed-loop aerials which, when used with a device known as a

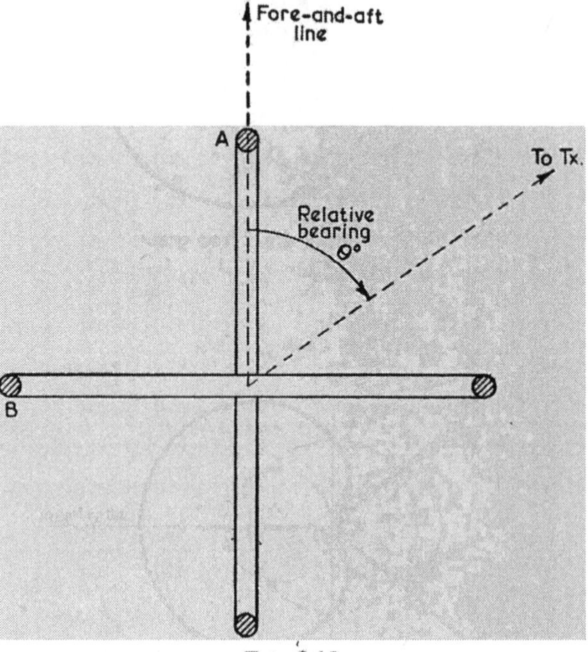

Fig. 2.10

goniometer, renders it possible to determine the bearing of a radio beacon or other distant transmitter.

The crossed-loop aerial system was first introduced by Bellini, Tosi and Marconi, and it is known as the B.T.M. aerial. The two loops are fixed such that they have a common vertical axis. One of the loops lies in the plane of the fore-and-aft line of the ship, and the other in the athwartship line.

The important feature of the B.T.M. aerial system is that the resultant of the e.m.f.s induced in the fore-and-aft and athwartships aerials is always constant regardless of the bearing of the transmitter. Consequently the maximum signal strength is solely dependent upon the intensity of the incoming energy and does not vary with the direction of the transmitter.

Fig. 2.10 represents a plan view of the B.T.M. aerials. Aerial A lies in the fore-and-aft line of the ship and aerial B lies in the athwartship line.

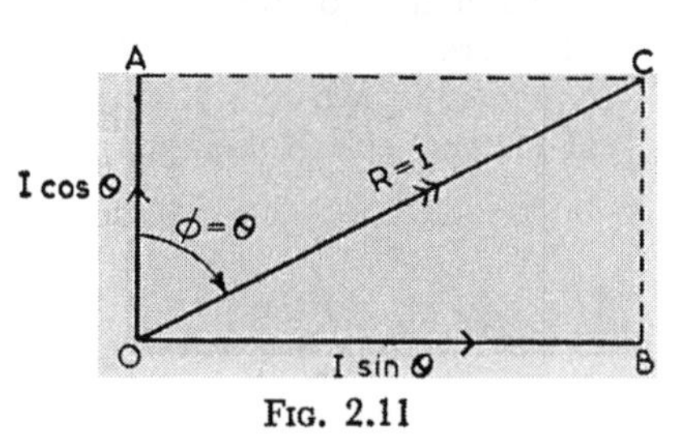

FIG. 2.11

Suppose a transmitter has a relative bearing of $\theta°$ as indicated, and suppose the maximum current induced in either loop is I. Now the current induced in loop A is maximum when θ is 0°, and zero when θ is 90°. It therefore varies as the cosine of θ. The current induced in aerial B is maximum when θ is 90° and zero when θ is 0°. It therefore varies as the sine of θ. Thus—

$$\text{Current induced in A} = \text{I} \cos \theta$$

$$\text{Current induced in B} = \text{I} \sin \theta$$

Because the loops are at right-angles to each other the resultant of the currents induced in aerials A and B is I, as indicated in Fig. 2.11, in which the induced currents are represented as vector quantities. The resultant R is found by means of the parallelogram law. By Pythagoras' theorem—

$$OC^2 = OA^2 + AC^2$$

Therefore

$$R^2 = I^2 \cos^2 \theta + I^2 \sin^2 \theta$$
$$= I^2(\cos^2 \theta + \sin^2 \theta)$$

Now

$$\cos^2 \theta + \sin^2 \theta = 1$$

Therefore

$$R^2 = I^2 \quad \text{and so} \quad R = I$$

Also, the angle ϕ, which the resultant makes with the vector $I \cos \theta$, is equal to θ because

$$\tan \phi = \frac{AC}{OA}$$
$$= \frac{I \sin \theta}{I \cos \theta}$$
$$= \tan \theta$$

Therefore

$$\phi = \theta$$

Hence it follows that the problem of finding θ—the relative bearing of a transmitter—resolves itself into the determination of the direction of the resultant field of those due to the induced e.m.f.s in the fore-and-aft and athwartship loop aerials. This is done by means of the goniometer.

Referring to Fig. 2.12, the ends of the loop aerials A and B are connected to two coils X and Y which are wound around a cylindrical former, coil X being wound in the vertical plane and coil Y

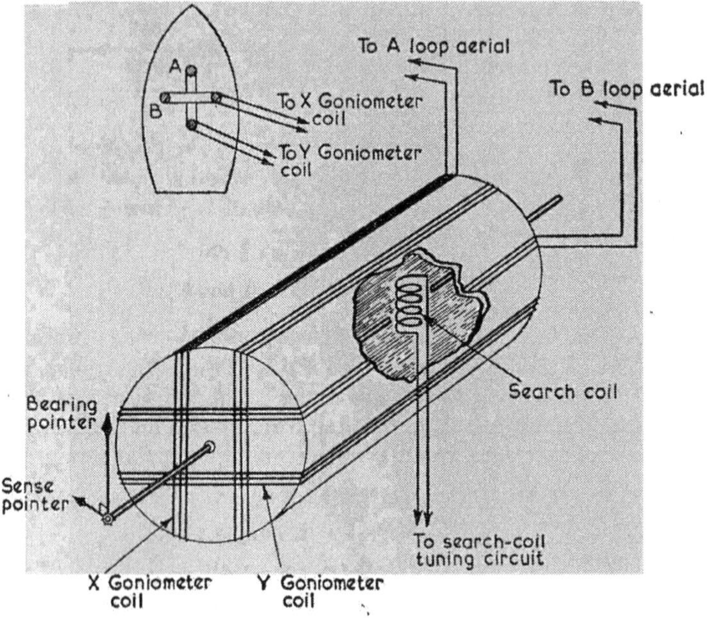

Fig. 2.12

in the horizontal plane. Along the axis of the cylindrical former lies a spindle on the end of which is fitted a bearing pointer which rides within a graduated compass card, the 0° and 180° marks of which are in line with the vertical coil X. Fitted to the spindle and lying within the coils X and Y is another coil, the axis of which lies at right-angles to the spindle. This is known as the *search coil*, for it is used to explore or search for the resultant magnetic field due to the currents in the X and Y coils, these currents being due to the e.m.f. induced in each of the loop aerials.

The loop aerials of the B.T.M. system are untuned aerials, the tuning to the transmitter frequency being achieved by means of the search-coil circuit.

The axis of the search coil lies in the same direction as the bearing pointer so that, when the bearing pointer indicates the bearing (or its reciprocal) of a transmitter, the plane of the search coil will lie in the plane of the resultant field due to the induced currents in the goniometer coils. Consequently no current will be induced in the search coil, and accordingly a zero or minimum signal will result.

Imagine a transmitter to bear 0° or 180°. The induced current in the fore-and-aft loop will be maximum and that in the athwartship loop will be zero. The induced current in the fore-and-aft loop will produce an alternating magnetic field within the goniometer coils, the direction of which will be horizontal. Now if the bearing pointer is turned by hand (in which case the search coil will also rotate) to the zero position, the search coil will then lie with its axis in the vertical plane, and no current will be induced in it, as the magnetic field due to the current in the vertical coil of the goniometer will be horizontal. Likewise if a transmitter bears 90° or 270°, when the bearing pointer is set to the zero position, the search coil will lie with its axis horizontal and will not be induced with current, as the magnetic field will lie in the plane of the coil, the field being a vertical one due to the alternating current in the horizontal coil of the goniometer.

A sense pointer is fitted to the goniometer spindle at 90° to the bearing pointer, and is used when sensing in the same way as with the rotating loop system.

To prevent changing magnetic fields, other than those due to the induced currents in the goniometer coils, from threading the search coil inside the goniometer, the former on which the coils are wound is surrounded by a cylinder of a metal which has a large amount of soft-iron property. This *magnetic screen*, as it is called, is normally made of mumetal.

The B.T.M. aerial system has an advantage over the rotating loop aerial on account of the mechanical difficulty involved in fitting a rotating loop.

Errors in Radio Direction Finding

The errors that may be involved in radio direction finding may be classified broadly into a two-fold division. Firstly are those errors which are due solely to the fact that the direction finder is fitted aboard a ship. It is possible to determine the magnitude and sign of such errors by the process of calibrating the direction finder. After calibration, a curve or table of errors is drawn up, such a record providing subsequently the correction to apply to any observed radio bearing. Alternatively, such errors may be eliminated by electrical or mechanical means. The second division of D.F. errors includes those which are caused by factors outside human control. The size and sign of such errors are usually unpredictable, and nothing can be done to eliminate them. Chief amongst these are errors due to "land effect" and "night effect."

Over the sea, radio energy travelling from a transmitter to a receiver moves along the shortest route, that is, along the great-circle track joining the transmitter and the receiver. As radio energy arrives at a ship, it induces alternating currents, not only in the direction finder aerials, but also in the hull and superstructures, and in all aerials and other conductors on the ship. Each of these alternating currents produces an alternating magnetic field which may thread the loop aerials of the direction finder, so causing possible error in the observed bearing.

Important amongst the alternating currents mentioned is that which is induced in an imagined large fore-and-aft vertical loop lying in the fore-and-aft axis of the ship. No current is induced in this *hull loop* (as it may be called), if the transmitter is abeam. If the transmitter is ahead or astern, maximum current is induced in the hull loop, and the resultant magnetic field due to the induction in the hull loop is athwartships; this would accordingly act with the field through the fore-and-aft loop aerial, and the true bearing of the transmitter would be the same as the observed bearing. If the transmitter is, however, in any direction other than right ahead, right astern, or abeam, the magnetic field due to induction in the hull loop will have a direction different from that of the magnetic field due to the direct energy arriving at the direction finder aerial. The resultant of these two magnetic fields will lie nearer the athwartship line of the ship than the magnetic field due to the direct energy, and accordingly the observed bearing of the transmitter will be too near the fore-and-aft line of the ship. Fig. 2.13 illustrates this when the transmitter lies on the starboard bow having a relative bearing of $\theta°$.

The magnetic field due to the induction in the hull loop is always athwartships, and its magnitude is proportional to the cosine of the relative bearing of the transmitter, that is, the field has maximum strength when the relative bearing is 0° or 180°, and zero strength when the relative bearing is 90° or 270°.

Quadrantal Error

If the D.F. aerial is located on the centre line of the ship with the mass of the ship symmetrical to port and starboard, and ahead and astern, the effect of induction in the hull loop, as described above, will be such that the true great-circle direction of a transmitter may be found by applying a correction to the observed radio bearing, towards the beam. Now an error (or correction), which changes its sign every 90°, as in the case described, is known as a *quadrantal error* (or correction). Quadrantal error in radio direction finding is due to the effect of radio energy on conductive objects in the vicinity of a loop aerial, which brings about a distortion of the magnetic (and electric) field which threads the loop aerial. The

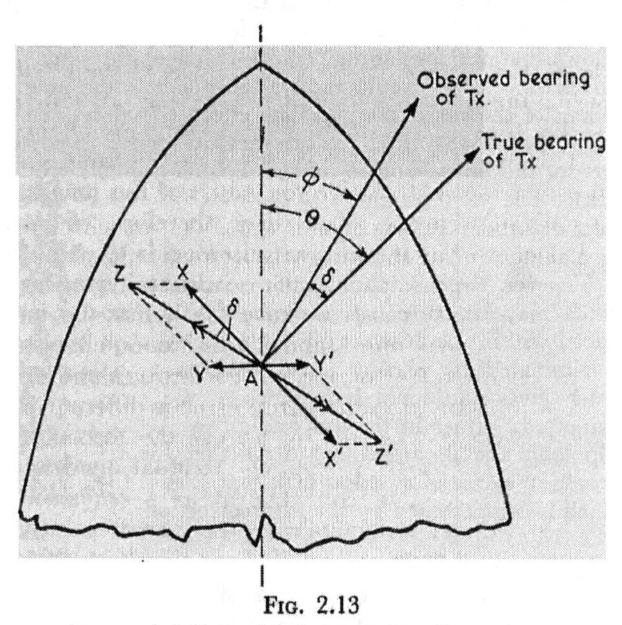

FIG. 2.13

XX' represents the magnetic field threading the loop aerial at A.
YY' represents the magnetic field threading the loop aerial due to the induction in the hull loop.

$$YY' \propto \cos\theta$$

ZZ' represents the resultant of XX' and YY'.
Angle δ represents the correction to apply to the observed bearing ϕ to obtain the true bearing θ.

greater the size and conductivity, and the closer the object to the loop aerial, the greater is the effect, for any given direction of a transmitter.

In order to correct for quadrantal error, a device known as a *quadrantal corrector* may be employed. In the case of the rotating loop aerial, the correction is sometimes made by means of an adjustable screw which carries an index which is set to the maximum quadrantal error, as determined by calibration, on a scale of degrees. Another method of correcting quadrantal error for rotating loop aerials is by means of a cam corrector which causes the bearing pointer to lag behind or lead on the loop aerial so that readings indicated by the bearing pointer are free of quadrantal error. The shape of the cam is such that any constant error, whether quadrantal or more complex, may be corrected automatically. The correct shape of the cam for this to be possible is determined from the initial calibration curve.

In the case of Bellini-Tosi aerials, compensation for a proportion of the quadrantal error is automatically made by having the fore-and-aft loop aerial smaller than the athwartship loop. The area of the loop aerial, it will be remembered, determines in part the magnitude of the induced current due to a changing magnetic field threading the loop. Because the radio energy arriving at a ship appears to be drawn towards the fore-and-aft line of the ship, it follows that the fore-and-aft loop—if it had the same size as the athwartship loop—would receive too much of the magnetic component of the radio energy. The effect, therefore, of placing the fore-and-aft loop within the athwartship loop is to compensate in part for this effect, and so part of the quadrantal error is compensated. The value of this *loop correction* (as it may be termed) is normally about 10° maximum, and if this amount happens to be the maximum quadrantal error, the loop correction alone is sufficient. If, however, the maximum quadrantal error is different from that which may be corrected solely by having the fore-and-aft loop placed within the athwartship loop, the residual quadrantal error is eliminated by placing a coil—known as a *calibration choke*—which gives additional inductance to the circuit in which it is connected: either in the fore-and-aft loop circuit or the athwartship loop circuit, according to whether the residual quadrantal error is in excess of or is less than that part of the quadrantal error which is compensated by the loop correction.

Semicircular Error

The vertical metal components of a ship's structure, such as pillars, masts, funnels, and samson posts, as well as standing and running rigging, act as aerials, and accordingly are induced with alternating currents which produce magnetic fields which may thread the loop aerials. Because of the symmetry of the vertical structures of the ship about the fore-and-aft line of the ship, the reradiated fields due to induction in vertical structures will be most intense in the athwartship direction, that is at right-angles to an imagined array of vertical aerials lying in the fore-and-aft line of the ship, and which is considered to replace the aerial effect of the vertical structures of the ship. The error due to induction in a ship's vertical structure on the fore-and-aft line is zero when the relative bearing of the transmitter is 0° or 180°, and maximum when the transmitter is abeam. The error due to this cause, therefore, changes its name every 180° and is thus described as a *semicircular error.* The sign of this error is dependent upon the location of the loop aerial relative to the disposition of vertical structures before and abaft it.

When the source of semicircular error is a permanent structure, the error produced by the source will be constant. The combined effects of such structures can be reduced or eliminated by a compensating device known as a *semicircular corrector.* The amount of semicircular error depends upon the structure of the ship and the disposition of masts, stays, vertical aerials, and so on. It may be determined by analysing the initial calibration curve. The amount of semicircular error is also dependent upon the frequency of the transmitter to which the direction finder is tuned, as this affects the magnitude of the induction in a particular vertical structure, being greatest when resonance occurs. This is a reason why, when calibrating a direction finder, a transmitter using a frequency within the radio beacon band should be chosen.

Figs. 2.14, 2.15, and 2.16, together with the relevant notes, serve to provide simple ideas on analysing calibration curves.

Fig. 2.14 illustrates a curve of quadrantal error in which the maximum error occurring on 045°, 135°, 225° and 315° is 10°.

Fig. 2.15 illustrates a curve of semicircular error in which the maximum error occurs on 090° and 270°, and has a value of 5°.

Fig. 2.16 represents the resultant of the two curves illustrated in Figs. 2.14 and 2.15.

The quadrantal error E_Q for any given relative bearing θ varies as sine 2θ. If therefore the maximum quadrantal error is 10°, as illustrated in Fig. 2.14,

$$E_Q = 10 \sin 2\theta$$

The semicircular error E_S for any given relative bearing θ varies as sin θ. If therefore the maximum semicircular error is 5°, as illustrated in Fig. 2.15,

$$E_S = 5 \sin \theta$$

The total error E_T for any relative bearing θ is therefore the sum of E_Q and E_S. That is—

$$E_T = E_Q + E_S$$

i.e.

$$E_T = 10 \sin 2\theta + 5 \sin \theta$$

This is the equation of the curve represented in Fig. 2.16.

In general

$$E_T = M \sin 2\theta + N \sin \theta$$

where M and N are the maximum values of quadrantal and semicircular errors respectively.

Assuming therefore that a calibration curve is due to a combination of quadrantal and semicircular errors—and this is readily

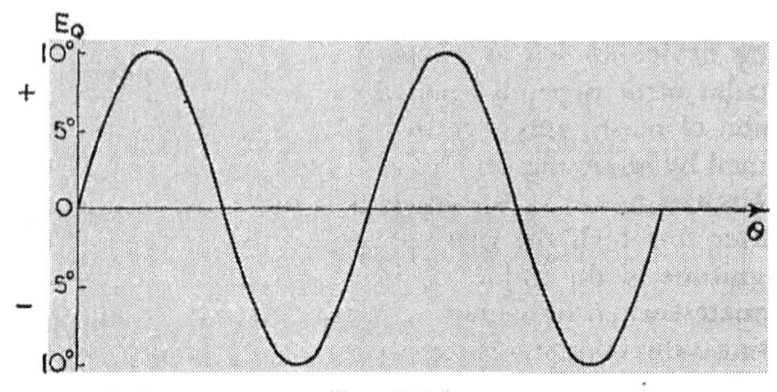

Fig. 2.14

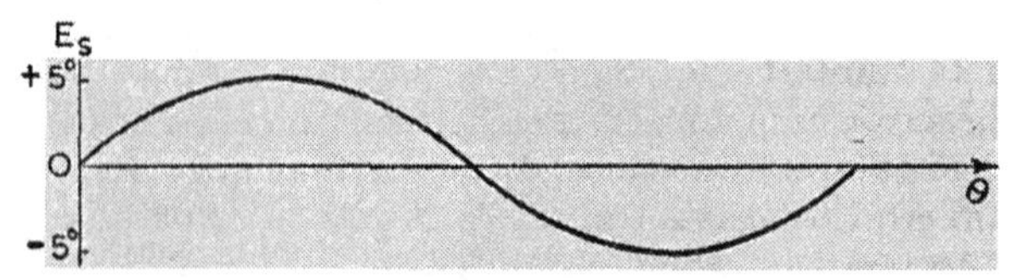

Fig. 2.15

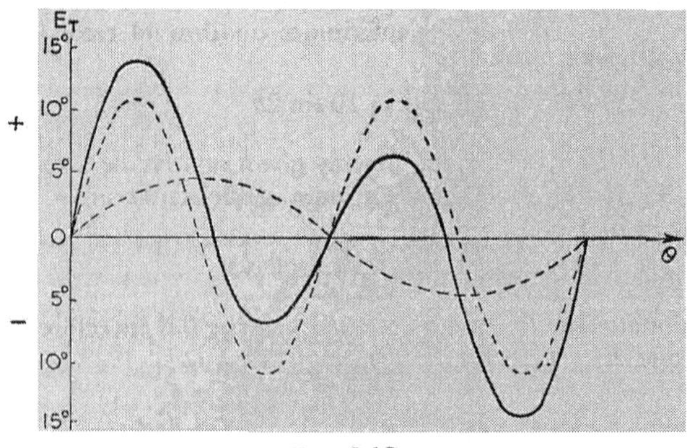

Fig. 2.16

determined by considering the shape of the calibration curve—the maximum values of M or N may be found algebraically by means of simultaneous equations involving values of E_T for two corresponding values of θ.

It frequently happens that vertical structures of the ship such as rigging, samson posts, or masts, together with the deck of the ship, form what is known as a *closed loop*. If the natural frequency of a closed loop is similar to the frequency of a transmitter to which the direction finder has been tuned, or if the frequency of the closed loop is a multiple or harmonic of the frequency of the transmitter, a strong magnetic field caused by induction in the closed loop may result. Should such a field thread the loop aerial of the direction finder, serious errors in the observed bearing of the transmitter may result.

In siting the loop aerial the question of closed loops in the vicinity should receive consideration. If it is unavoidable in having to site the loop aerial within a closed loop, efficient insulators should be fitted to eliminate the possibility of induction in the loop. Theoretically one insulator in such a loop is sufficient, but it is desirable to fit several so that the loop is broken up into reasonably short lengths. This applies particularly to triatic stays and steel whistle lanyards which are located in close proximity to the loop aerial.

The loop aerial should be located well clear of funnels and ventilators when practicable. Movable ventilators near to a loop aerial present a difficult problem to those who have the task of fitting a loop aerial.

When using a radio direction finder, the ship's main aerial should be broken. The influence of an unbroken main aerial is to increase the apparent deflexion of the incoming radio energy into the ship's fore-and-aft line. This influence is greatest should the main aerial be tuned to the same frequency as the incoming signal.

The two principal errors in radio direction finding which fall outside human control are errors due to what are referred to as *land effect* and *night effect*.

Land Effect

When light—which is a form of electromagnetic energy—passes from one medium to another of different optical density its velocity changes. As a consequence of this its path is bent at the common surface of contact between the two media. The angle through which the path is bent is known as the *angle of refraction*, and its value is dependent upon the optical densities of the two media and the angle made by the path of the light with the normal to the common surface of contact.

A similar effect to the refraction of light is the bending of a train of radio energy as it passes the common boundary of two surfaces which have different electrical conductivities. When radio energy travels between a transmitter and a receiver, such that part of its path is over land and the other part is over sea, its direction changes at the coast by an amount dependent upon the angle at which the energy crosses the coastline. The velocity of radio energy over the sea may be as much as a twentieth greater than its velocity over

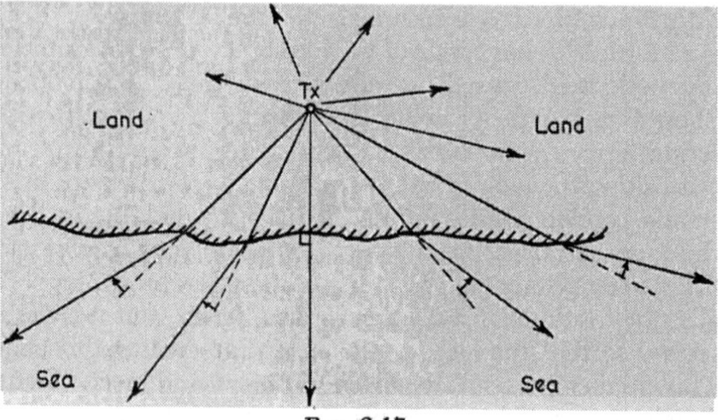

FIG. 2.17

land. As with light, bending is always away from the normal in the case of energy travelling into a medium in which its velocity is increased.

The change in direction of radio energy as it crosses the land edge is referred to as *coastal refraction*. On crossing the coast after travelling over land, the refraction is towards the edge of the land.

When the line joining a transmitter and receiver is normal to the coast, that is to say, when the line joining a transmitter and receiver crosses the coast at an angle of 90°, coastal refraction is zero. The angle of coastal refraction increases as the line joining the transmitter and receiver makes a decreasing angle with the coast. Fig. 2.17 illustrates this phenomenon. Tx represents a transmitter and the full lines represent paths of rays of radio energy emitted from the transmitter. It will be noticed from Fig. 2.17 that the coastal refraction is towards the land, so that a false position line laid down on a chart is always too near the coast which has been crossed by the received energy. Fig. 2.18 illustrates this point.

Referring to Fig. 2.18, suppose S represents the position of a ship on which the radio bearings of transmitters represented by Tx_1 and Tx_2 are observed. The false position line resulting from the radio bearing of Tx_1 is P_1L_1, and that obtained from Tx_2 is P_2L_2. Fig. 2.18 serves to illustrate that the greater the distance of the transmitter from the coast, the greater will be the displacement of the false position line from the true position of the ship.

A graphical representation of coastal refraction is given in Fig. 2.19, in which the x-axis is graduated in terms of the angle between

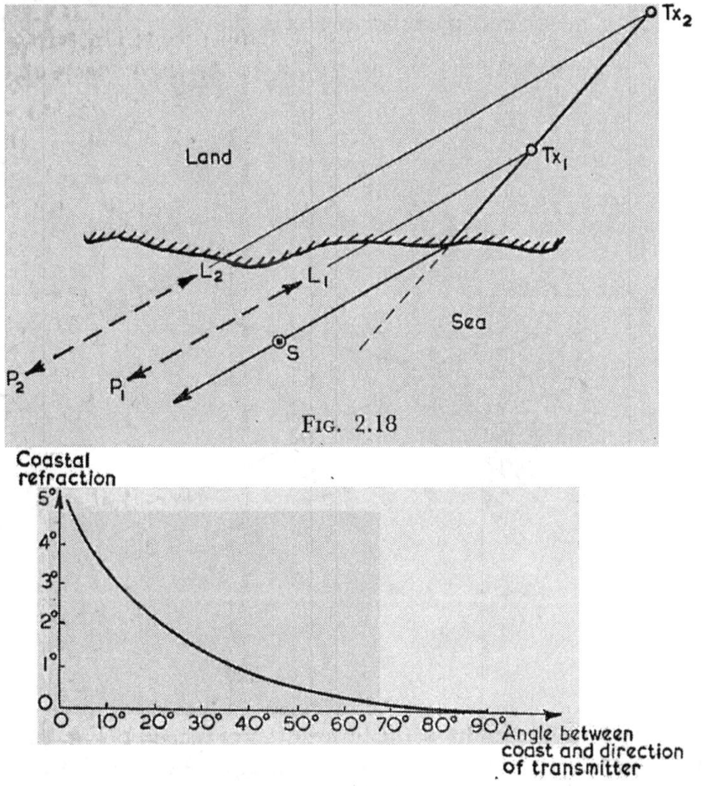

FIG. 2.18

FIG. 2.19

the coastline and the line joining the transmitter and receiver; and the y-axis is graduated in degrees of coastal refraction. As shown, coastal refraction may be as much as 5° when the angle between the coastline and the direction of the transmitter is very small. When this angle is greater than about 50° the angle of coastal refraction is negligible. Therefore, when choosing a radio beacon or other land-based transmitter for the purpose of obtaining a radio position line, the prudent navigator should consider these important points.

The refraction of light depends in part upon the frequency of the light energy, being least for light at the red or low-frequency end of the visible spectrum, and greatest for light at the blue or high-frequency end of the spectrum. A similar effect is to be found in the case of coastal refraction, the angle varying as the frequency of the radio energy for values of frequency above about 150 kilocycles per second.

Errors in radio bearings, in addition to those caused by refraction, may be caused by reflexions of radio energy from coasts or high

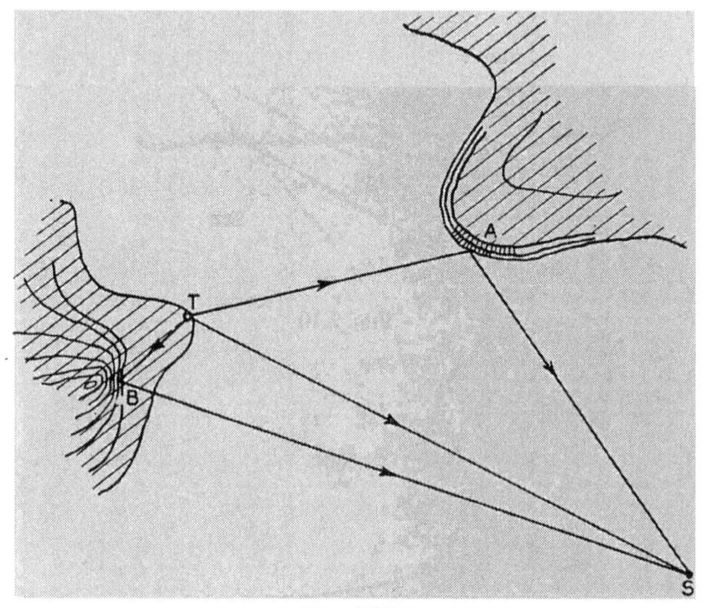

Fig. 2.20

land in the vicinity of the transmitter or receiver. Fig. 2.20 illustrates the effect of reflexion of wireless energy. A ship at S, whose direction finder is tuned to station T may receive, in addition to the direct energy from T, reflected energy from the cliff at A and/or from the high precipitous mountain at B. The result of receiving indirect as well as direct energy may be an error in the observed bearing of as much as 6° or 7°. Under conditions similar to those depicted in Fig. 2.20 radio bearings should not be relied on implicitly.

In general, best radio bearings are obtained from radio beacons located on light vessels well clear of the land, providing that no land intervenes between the beacon and the ship; or from beacons located on the land edge—on lighthouses for example—providing that no high land lies close behind the site.

When calibrating a direction finder it is important to choose a suitable radio beacon and to position the ship so that the effects of coastal refraction or coast reflexion are kept to a minimum.

Night Effect

The determination of the bearing of a transmitter by means of a loop aerial, as described earlier, is based on the assumption that the radio energy arriving at the loop aerial is travelling parallel to the Earth's surface. In other words, the radio energy is horizontally polarized. This is the case when the loop aerial is being affected solely by the ground-wave radiation from a transmitter. If, however, sky-wave radiation is being received, the loop aerial is being affected by radio waves which are striking it obliquely, that is to say, the radio waves being received are not horizontally polarized.

Imagine a single rotating loop aerial to be set up so that its vertical plane is at right-angles to the direction of a transmitter to which the aerial has been tuned. The alternating magnetic field due to ground-wave radiation will not thread the loop, and so no e.m.f. will be induced in the loop, and accordingly a zero signal will result. If, however, the plane of the loop aerial is at right-angles to the direction of the transmitter, and sky-wave radiation is being received, the loop aerial will be threaded by the changing magnetic field associated with the radio energy, if the plane of polarization of the radio waves has been rotated at the point in the ionosphere at which the waves were reflected back to the Earth.

If the loop aerial is threaded by a changing magnetic field, even though the plane of the loop aerial is at right-angles to the direction of the transmitter, an audible signal will result. The zero-signal position of the loop will not, under these circumstances, render possible the determination of the direction of the transmitter. Hence, when receiving sky waves there is a possibility of an error in the observed radio bearing of a transmitter. It must be noted that this error, if it exists, is not due to the energy reaching the loop aerial along a path which is oblique to the horizon, but is due to the radio energy, after reflexion at the ionosphere, not travelling along the same plane as its path before reflexion took place.

In general, during the daytime, the ionosphere acts as a horizontal reflector of radio energy, in which case incident and reflected radio energy lie in the same plane. This is illustrated in Fig. 2.21 in which it will be noted that the plane in which the incident and reflected rays lie is vertical. If for any reason the ionosphere does not act as a horizontal reflector (in which case it may be imagined that the ionosphere is tilted to the horizontal), the incident and reflected energy will no longer lie in a vertical plane. This is illustrated in

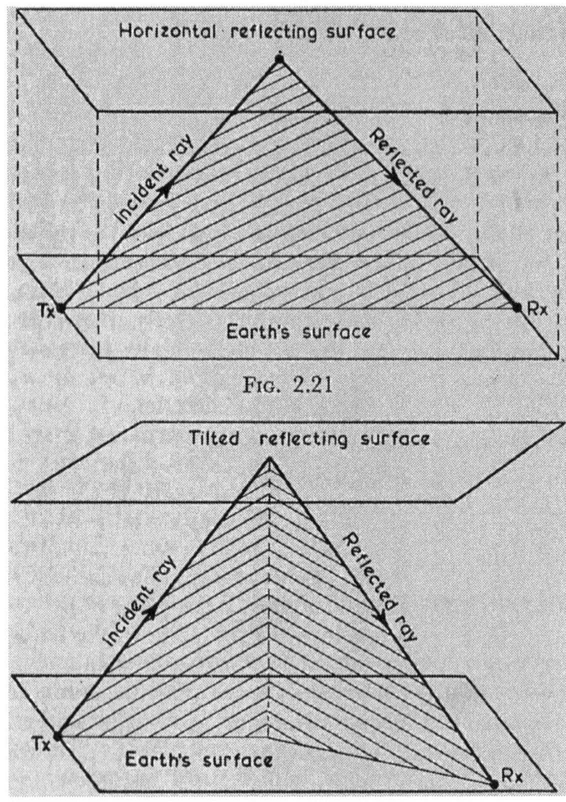

Fig. 2.21

Fig. 2.22

Fig. 2.22, in which it will be noted that the directions of the vertical planes of the incident and reflected rays are different.

During daylight hours sky-wave radiation is relatively weak for radio-beacon frequencies. But between sunset and sunrise sky-wave radiation is strong, and, at ranges at which both ground and sky waves are being received outside the skip distance, the greater proportion of the signal received is normally due to sky-wave radiation. When receiving sky waves there is always a possibility of error arising in observed radio bearings. The liability of such error occurring is a function of the ratio of the values of the sky- and ground-wave radiation being received. This ratio is greater during dark hours than what it is during daylight, and so errors in bearings due to sky-wave reception are more frequent during the night time than during daytime; hence the name *night effect.*

The ionosphere is normally in a particularly disturbed state near the times of sunset and sunrise (the Sun being the principal agent of ionization of the atmosphere), and at or near these times error due to the phenomenon of night effect is most pronounced and common. During the daytime, when the ratio between sky- and ground-wave radiation being received is small, good results from observations by a radio direction finder are to be expected—bearings to within about 2° of the truth up to a range of about 100 miles are general, if the instrument or observation are not subject to other direction finding errors. During the night-time, the reliable range is reduced to within the ground range to about 25 miles. Outside this range, the greater proportion of received energy is due to sky-wave radiation, and consequently there is every possibility of error due to night effect. Within an hour or so of the times of sunset and sunrise, on account of error due to night effect, reliance can seldom be placed on radio bearings.

The term night effect, as applied to the phenomenon described above, is a misnomer, as disturbances in the ionosphere which result in the same phenomenon may occur during daytime. Sunspot activity and severe electrical storms within the troposphere often give rise to night effect during daylight hours.

Generally the presence of night effect is accompanied by fluctuations in the signal strength at the receiver, known as *fading.* Another possible indication of night effect is a variation in the apparent direction of the transmitter, a phenomenon referred to as a *wandering zero.* Fortunately, the presence of night effect is easily discovered by a trained observer. Whenever night effect is known or suspected to be present, radio bearings should not be relied upon.

Part 3

The Practice of Radio Direction Finding

THE ship's radio direction finder, if it is to be a useful and effective instrumental aid to navigation, must be properly installed and maintained, and fully and accurately calibrated. Furthermore it should be operated by properly trained and intelligent officers.

This part treats of the practical aspect of radio direction finding and includes information about some current direction finding equipments.

The loop aerials of the direction finder should be placed on the centre line of the ship, such that the mass of the ship is symmetrical about it. In choosing the site for the aerial, attention should be paid to the length of cable between the loop serials and the instrument: the length should be kept to a minimum. The site should be free from aerial leads, rigging, lightning conductors, and large metal masses; railings and stanchions should be carefully bonded and earthed; and standing rigging in the neighbourhood of the loop aerials should be broken up by suitable insulators. The best practical position is on the upper bridge of the ship. When the aerial is placed high above the hull of the ship, errors due to the ship itself are minimized.

Calibration of Radio Direction Finders

The calibration of a ship's radio direction finder involves the comparison of simultaneous visual and radio bearings of a distant transmitter, on relative bearings around the compass at intervals which should not exceed 5°. Before conducting a calibration procedure, the ship should be in her normal sea-going condition in respect of her standing and running rigging, and all portable metal gear in proximity to the direction finder. The standing rigging should be set up, and the running rigging dismantled or secured according to the practice in the ship. The derricks or cranes should be in their stowed position, and likewise all metal lockers and similar portable gear should be secured in their sea places. Before commencing to calibrate, the instrument should be checked for mechanical and electrical performance.

The calibration of a direction finder may be carried out by one of two methods. The ship may be anchored or moored and simultaneous visual and radio observations of a transmitter are made, the transmitter being carried in a tug, launch or other craft, which steams slowly around the ship. The observations are made when the relative bearing of the craft is a multiple of 5°, and the results are recorded either graphically or in a tabular form.

The alternative method of calibrating a direction finder is to swing the ship in relation to a fixed radio station—preferably a radio beacon or a special calibrating station. When swinging the ship for this purpose, the ship should be steadied on courses at intervals of not more than 5°, and the compass card should be quite steady on each heading before making the observations.

The calibration data should include details of all other aerials in the vicinity of the loop aerial, and whether they are isolated or connected, and also the positions of any movable deck structures, so that the same conditions may be made to obtain when using the instrument as an aid to navigation. Calibration must of course be carried out by two persons, one to observe the visual bearings and the other to observe the radio bearings. It should be carried out well clear of the land and other transmitters. The calibration table or curve should be fully verified by means of check-bearings in each period of twelve months. If at any time check-bearings reveal discrepancies from the data obtained at the last calibration, the instrument should be used with extreme caution and should be recalibrated at the earliest opportunity.

A suitable and adequate power supply should be available for the direction finder. If this is provided by means of a battery of accumulators they should be tested daily by a voltmeter and monthly by a hydrometer. The accumulators should never be allowed to remain in an uncharged condition, and they should never be overcharged.

In ships which are required to be provided with a direction finder, the following records* must be kept on board in a place accessible to any officer who may operate the instrument—

1. A list or diagram indicating the position and condition at the last occasion on which the direction finder was calibrated of the aerials on board the ship and of all movable structures which might affect the accuracy of the direction finder.

2. The calibration tables and curves which were prepared as a result of the last calibration.

* See Appendix, page 105.

3. A certificate of calibration in the form specified in the Rules signed by the persons who made the last calibration.

4. A record, in the form specified in the Rules, of all check-bearings taken for verification purposes, and of any other information which might be useful in showing whether or not the equipment gives satisfactory performance.

5. Where applicable, a record of the battery tests.

Masters of ships fitted with direction finders should be familiar with M.O.T.C.A. Notice Number M.402, which draws attention to all concerned with the fitting and use of direction finders on ships, to the necessity of ensuring that such equipment is properly installed, maintained, and fully and accurately calibrated.

Radio Direction Finding Stations

Radio direction finding stations provide services by means of which ships at sea are able to determine a radio position line or a radio position. The Service Details of all radio D.F. stations are given in *The Admiralty List of Radio Signals*, Volume 2. Typical sets of service details are those for the radio D.F. stations of Land's End and Georgetown,* given below.

1011 Land's End (GLD). Receiver: 50° 07′ 09″ N. 5° 40′ 40″ W.
Transmitter: 50° 07′ 04″ N. 5° 40′ 05″ W.

FREQUENCY: (A) 410. A2. [0800—2300.]
500. A2.
(B) 500, 410.
(C) 500, 410. A2. 5·0 kW.

SECTORS: Calibrated: On 410 kc/s: 001°—057°, 067°—360°.
On 500 kc/s: 001°—058°, 067°—360°.

1843 Demerara (Georgetown) (VRY).
Receiver: 6° 49′ 32″ N. 58° 08′ 49″ W.
Transmitter: 6° 49′ 30″ N. 58° 08′ 50″ W.

FREQUENCY: (A) 500. A1, A2, B.
(B) 410 (see NOTE). A1, A2, B.
(C) 500. A1, A2. 0·75 kW.

NOTE: Ships unable to transmit on 410 kc/s may use 500 kc/s or 425 kc/s.

HOURS: 0945–0345. If required at other times, station must be notified before 0345.

SECTORS: Calibrated: 000°—360°.
Approximate: 076°—299°

PROCEDURE: As for British D/F Stations.

CHARGE: Five shillings per bearing.

* Reproduced by kind permission of H.M.S.O.

Each station, it will be noticed, has an index number, e.g. 1011 for Land's End. Following the index number is the name of the station, and this is followed by a three letter identification or call signal. The positions of the receiver and transmitter are then given, and also three frequencies labelled (A), (B), and (C), to be used as described hereunder. Normally, service is continuous throughout the day, but if it is not, as in the case of Georgetown, the hours of working are given in the service details. Radio D.F. stations do not necessarily give all round coverage, and the sectors centred at the stations within which ships may expect to obtain reasonably good bearings are those for which the receiver at the station has been calibrated. The calibrated sectors for Land's End D.F. station are 001° to 057° and 067° to 360° when using a frequency of 410 kc/s; and 001° to 058° and 067° to 360° when using 500 kc/s. Notice that the calibrated sectors may be slightly different for different frequencies. The limits of calibrated sectors are true bearings and are given clockwise from the station. D.F. bearings are to be considered reliable only when they fall within a calibrated arc. Before employing the services of a radio D.F. station, it is essential, therefore, if reliance is to be placed on the radio bearing from the station, for the ship to be located within an arc of good bearings.

When it is necessary to obtain a radio position line by means of the facilities offered by a radio D.F. station, the ship calls the station using the frequency designated (A) in the service details. This frequency is sometimes known as the listening frequency, as it is the frequency to which the operator at the station has his receiver tuned. The operator at the D.F. station, on receiving the ship's call, by means of the appropriate service abbreviations (QTE) in this case), requests the ship to transmit a signal using the frequency designated (B) in the service details, so that the operator at the station may determine by means of his direction finding equipment, the radio bearing of the ship. For radio D.F. stations in Great Britain the transmitted signal is comprised of two dashes of 10 sec each followed by the ship's call sign. For German stations it is the ship's call sign transmitted for a period of 1 to 2 min. The appropriate signal is described for all countries in *The Admiralty List of Radio Signals*, Volume 2.

When the radio bearing has been obtained, the operator at the station calls the ship on the frequency designated (C), and gives: the true bearing of the ship from the station; the class of the bearing; and the time of the observation. The ship repeats this information, and the transmitting station confirms that the repeated signal is correct or states that it is not.

According to the estimation of the degree of accuracy made by the operator at the station, radio bearings are classed as follows—

Class A—Bearings accurate to $\pm 2°$
Class B—Bearings accurate to $\pm 5°$
Class C—Bearings accurate to $\pm 10°$

Before laying down a radio bearing as a position line on a Mercator chart, it must be remembered that the radio bearing is the great-circle bearing of the ship from the station. The half-convergency correction must therefore be applied equatorwards, in order to obtain the rhumb-line bearing. The half-convergency correction, which is due to the convergency of the meridians polewards, may be found from a table of corrections to radio bearings as given in most sets of nautical table; or by means of a half-convergency diagram, one type of a variety of which appears in *The Admiralty List of Radio Signals*, Volume 2.

Half convergency may be calculated from the formula—

half convergency = half d.long. between ship and station multiplied by the sine of the middle latitude of the latitudes of the ship and station.

That is

$$\tfrac{1}{2} \text{ convergency} = \tfrac{1}{2}\text{d.long. sin.mid.lat.}$$

The correction may thus be found quickly by means of the Traverse Table. It will be noticed from the formula that the greater the d.long. between the ship and the station, and the higher the middle latitude, the greater will be the half convergency.

To obtain a *position*, or *radio fix*, through the medium of radio D.F. stations, the ship calls the station which controls a group of radio D.F. stations, and makes her request by means of the appropriate service abbreviation—QTF in this case. On receipt of the signal from the station requesting the ship to transmit, the ship transmits, on the appropriate frequency, a signal by means of which three widely separated radio stations obtain radio bearings of the ship. Available at the control station is a chart of the region on which the radio D.F. stations are plotted, each with a compass rose around it. The operator at the control station lays down on this chart the three uncorrected bearings as position lines. Having done this the approximate position of the ship is known and the half-convergency correction for each of the three position lines may be determined. The position lines are then adjusted for this correction, and a more exact estimation of the ship's position is then found.

When the ship's position has been found, the ship is called and informed of the position in terms of latitude and longitude. Also given is the class of the position and the time of the observation. Positions are classified by the operator at a British control station as follows—

Class A—Position accurate by estimation to within 5 nautical miles.
Class B—Position accurate by estimation to within 20 miles.
Class C—Position accurate by estimation to within 50 miles.

To ensure best results when using the services provided by radio D.F. stations, the signal transmitted by the ship in order to allow radio bearings to be taken should be clear and strong, steady in note and strength, and transmitted on the correct frequency.

In general, bearings taken by radio D.F. stations are accurate to within 2° of the truth. Observations are, however, subject to errors (described in Part 2), and therefore a prudent navigator should not place undue reliance on the results of such observations. It must be remembered also that no responsibility for consequences of bearings or positions being inaccurate will be taken by the authority which controls any radio direction finding service.

Radio Beacons

Ships which are fitted with radio direction equipment may obtain a *radio bearing* by tuning to any station which is transmitting on a frequency within the range of that of the receiver. The accuracy of a radio bearing is dependent—amongst other things—on the frequency of the transmission. It has been found that best results are normally obtained on frequencies within 285–315 kc/s. This part of the radio-frequency band is referred to as the *radio beacon band.* Accordingly, the numerous radio beacons, which have been established as aids to navigation, operate in the main on frequencies which fall within this range. Hence for radio direction finding, using transmitting stations other than radio beacons, undue reliance should not be placed on results for navigational purposes unless the receiver has been calibrated for the frequency used.

Radio beacons normally operate during conditions of poor visibility, but clear-weather transmissions are also made by many beacons. A radio beacon transmits, on a specified frequency, a characteristic signal by means of which the station may be identified. The service details of radio beacons are to be found in *The Admiralty List of Radio Signals,* Volume 2. The service details for any radio beacon include the name and index number of the station; its position and the frequency used in the transmission; the characteristic and its period; details of clear weather transmissions and fog service; and any other relevant information.

Typical of service details of radio beacons are those for Algiers in the Mediterranean, and Lundy Island North Lighthouse in the Bristol Channel, given below.*

2347H Algiers. 36° 46′ 50″ N. 3° 04′ 27″ E.

FREQUENCY: 305·7. A2. RANGE: 20 miles.

CHARACTERISTIC: Period 30 sec.

AL (· — · — · ·) twice	5 sec
Long dash (——)	3 ,,
Silent	2 ,,
30 dots (· · · etc.)	18 ,,
Silent	2 ,,
Period	30 sec

CLEAR WEATHER TRANSMISSIONS: NONE.

FOG SERVICE: Continuous.

AIR FOG SIGNAL, every 30 sec.

Silent	4 ,,
Blast 3 sec, silent 3 sec, blast 1 sec	7 ,,
Silent	19 ,,
Period	30 sec

SYNCHRONIZATION: The 1-sec blast of the air fog signal commences at the same time as the first of the 30 successive dots of the radio beacon. The number of dots received before the 1-sec blast of the air fog signal gives distance from the station, each dot corresponding to a distance of 1 cable.

2012 Lundy I. North Lt. Ho. 51° 12′ 03″ N. 4° 40′ 34″ W.

FREQUENCY: 296·5. A2. RANGE: 50 miles. SEQUENCE No.: V.

CHARACTERISTIC: Period 6 min.

NL (— · · — · ·) 4 times	19·6 sec	(0 min 53·7 sec)
Long dash (——)	25·0 ,,	
NL (— · · — · ·) twice	9·1 ,,	
Silent	306·3 ,,	(5 min 6·3 sec)
Period	360·0 sec (6 min)	

BEACON SERVICE: Continuous, commencing at 4 min past each hour

REMARKS: Grouped with South Bishop Lt. Ho. (No. 2017), Skerries Lt. Ho. (No. 2018), Cregneish (No. 2021), Kish Lt. V. (No. 2067) and Tuskar Rock Lt. Ho. (No. 2068).

* Reproduced by kind permission of H.M.S.O.

The characteristic is the cycle of transmission of a radio beacon in a given period of time. The range given in the service details is normally the ground range in nautical miles. When two values are given, thus 150/100, the first is the range by day and the second the range by night. In some instances the power of the beacon is denoted by a classification letter, A,B,C, or D. The useful ranges corresponding to these classes are: 200, 100, 20, and 10 miles respectively.

Radio beacon service in Europe and North Africa is organized on a rational basis. All beacons within this area operate on frequencies within the radio beacon band (285–315 kc/s), and their identification signals are of two letters, chosen to facilitate easy recognition, such as KR for Mull of Kintyre, or BL for Butt of Lewis, etc.

Normally each radio beacon transmits its signal at a speed of between six and ten words a minute, and the characteristic signal has been standardized for areas north and south of the parallel of 46° N—the period being 60 sec for beacons north of latitude 46° N, and 120 sec for beacons situated south of latitude 46° N.

To provide facilities for observing cross-bearings, groups of beacons have been formed, each group comprising not more than six radio beacons, all of which transmit their signals on the same frequency. The complete group transmission is normally of 6 min duration, and during this period each station will transmit its characteristic a number of times dependent upon the number of stations in the group: once when the number is six; twice when three, and three times when the group consists of two stations only. Lundy Island North Light House beacon—it will be noticed—is number five in the sequence of signals transmitted by the group comprising Lundy I. N.Lt.ho., South Bishop Lt.ho., Skerries Lt.ho., Cregneish, Kish Light vessel, and Tuskar Rock Lt.ho.

Many radio beacons in Great Britain and Northern Ireland provide a calibration service which may be either a *routine* or a *request* service. The routine service involves continuous service from one hour after sunrise to one hour before sunset, or for one hour in the morning and afternoon in clear weather only. The request service is available between one hour after sunrise to one hour before sunset, when the station is not otherwise engaged. Details of these services, and the corresponding services provided by other maritime nations, are given in *The Admiralty List of Radio Stations*, Volume 2.

QTG Service

If a ship which is equipped with radio direction gear is not located conveniently in relation to radio beacons, but is in a suitable position with respect to a coast radio station, and if a radio bearing of such a station would be desirable, the ship may take advantage of a service provided by numerous coast radio stations, which will transmit for radio direction finding purposes, on request. Coast radio stations providing this service—known as the *QTG service*—are listed at the end of each geographical section in *The Admiralty List of Radio Signals*, Volume 2, but fuller details, giving hours of service and frequencies employed, are to be found in Volume 1 of the same publication.

When using the QTG service, the coast station must be called in the normal way. To facilitate the determination of a radio bearing by the ship's direction finder, the signal QTG is made by the ship. This signal signifies "Will you send your call sign for 50 sec followed by a dash of 10 sec on a frequency of .... kc/s, in order that I may take your bearing?"

Numerous aircraft radio navigational aids have been established on fixed sites and are operated on regular schedules. Some of these aids may be useful for marine navigational purposes. Details of certain selected stations are given at the end of each geographical section in *The Admiralty List of Radio Signals*, Volume 2. Air radio beacons are operated in a manner similar to the operation of marine radio beacons. The characteristic usually consists of the station's identification signal combined with a long tuning dash.

Radio ranges are used primarily for homing aircraft. A continuous signal is to be heard on four narrow bearing sectors—known as *range legs*—and the letters A (.–) or N (–.) in the sectors between the range legs. At intervals, usually of a minute, the transmission is interrupted by the station's identification signal in morse code.

Certain radio beacons provide the means whereby the range of the beacon may be determined by a ship, whose direction finder is tuned to the frequency of the station. This is possible by the synchronization of radio and sound signals. The sound may be emitted in air or water. During the radio transmission, at a particular phase, the beginning of a sound signal is made. Details are given under the heading Synchronization, in the service details of the radio beacon.

The time interval between the instants of receipt of the radio and sound signals at the ship is a function of the range. Knowing the speed of sound in water and air, it is a simple matter to determine the range.

$$\text{Speed of sound in air} = 1{,}100 \text{ ft/sec}$$

$$= \frac{1{,}100}{6{,}080} \text{ miles/sec}$$

$$\text{If the time interval} = t \text{ sec}$$

$$\text{then the range} = t \times \frac{1{,}100}{6{,}080} \text{ miles}$$

$$0{\cdot}18t \text{ miles}$$

$$\text{Speed of sound in water} = 4{,}800 \text{ ft/sec approx.}$$

$$= \frac{4{,}800}{6{,}080} \text{ miles/sec}$$

$$\text{If the time interval} = t \text{ sec}$$

$$\text{then the range} = t \times \frac{4{,}800}{6{,}080} \text{ miles}$$

$$= 0{\cdot}8t \text{ miles}$$

Thus, for a submarine signal, multiply the interval in seconds by 0·8, and for an air signal multiply the interval in seconds by 0·18, in order to determine the range in miles.

In many cases it is not necessary to perform a calculation, as in the transmission of regularly spaced dots or dashes, the period between which is such that the number of signals transmitted, in the interval between the times of receipt of the radio and sound signals, is equal to the range in miles or fractions of a mile.

Although the speed of sound in air varies with temperature, pressure, and humidity, and in water with salinity, density, and temperature, the figures used to compute the range, namely 0·8 and 0·18, are sufficiently accurate for practical purposes.

In ships which are not fitted with hydrophones, submarine signals may be heard from an enclosed compartment below the ship's waterline—a fore peak store or tank is suitable, provided the ship is not making way through the water.

Special arrangements apply to some radio beacons, for example, Cloch Point Light house in the Firth of Clyde, by means of which the range may be found. In the case of Cloch Point Light house, the range in cables is spoken by radio telephony, on a frequency of 301·1 kc/s. A similar arrangement applies to Little Cumbrae Light house, also on the Firth of Clyde.

Ocean weather ships in the North Atlantic and North Pacific are equipped with radio beacons, which, because weather ships normally remain when on station within a small area, may be used with advantage for the purposes of position finding. The characteristic signal of the weather ship radio beacon consists of four

OCEAN WEATHER SHIPS

Position Indicating Grid

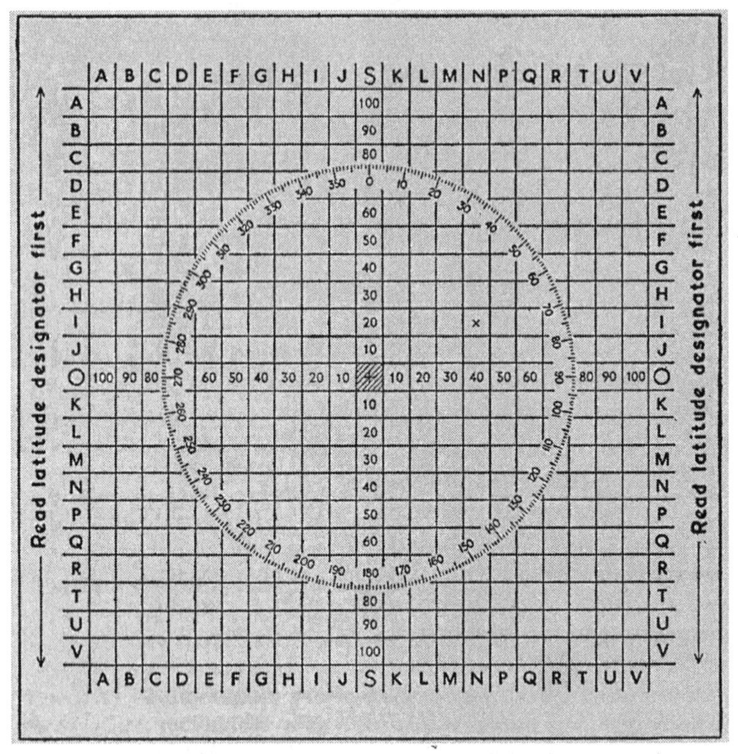

Fig. 3.1. Position Indicating Grid for Ocean Weather Ships

(Reproduced by courtesy of H.M.S.O.)

Note. The grid lines are 10 nautical miles apart. The latitude designator is always given first.

The centre of the grid is the geographical position assigned to, and normally occupied by, the station.

The latitude and longitude designators are always added to the identification signal of the ship's radio beacon. For example, if the identification signal is QJ and the vessel is on station (i.e., within the ten mile square at the centre of the grid) the transmitted characteristic signal will be QJOS. If the vessel is off station (but within the grid limits) the latitude and longitude designators of the square actually occupied will be added to the identification signal. (The centre of each grid square should be considered the location of the vessel for all computations, thus giving a maximum possible error of 7·5 miles and an average probable error of 2·5 miles.)

Example: If the ship's actual location is at the point marked × on the grid above and the identification signal is QJ, the transmitted radio-beacon characteristic would be QJIN. It will be seen that in this position the ship would be about 64° true, 45 miles from its assigned position.

letters, the first two of which are the identification signal of the station, and the second two of which indicate the position of the weather ship in relation to its assigned geographical position. This is shown in Fig. 3.1 which is a reproduction of page 325 of *The Admiralty List of Radio Signals*, Volume 2.

Prior to fixing a ship's position by means of radio bearings, suitably placed radio beacons must be selected. In European waters, it is useful to choose stations whose transmissions are coordinated, as this facilitates the rapid determination of two or more bearings necessary for fixing by cross-bearings. In choosing radio beacons, assistance is often afforded by referring to the index maps which are contained at the end of *The Admiralty List of Radio Signals*, Volume 2. These index maps cover, collectively, the whole of the navigable coastal regions of the Earth. Marine radio beacons are plotted on the index maps in red, and aeronautical radio beacons are plotted in blue. After deciding on suitable beacons, the names of which are shown on the relevant index map, the index numbers of the stations are found from the alphabetical index, which is to be found in *The Admiralty List of Radio Signals*, Volume 2, immediately before the index maps.

Also to be found near the end of *The Admiralty List of Radio Signals*, Volume 2, are: a conversion table for converting frequency into wavelength; a diagram for determining half-convergency corrections; and a list of the signals from the radio officer's Q code, which may be used in the practice of radio direction finding.

The Marconi Lodestone Direction Finder

The Lodestone direction finder has a frequency range of 250 to 550 kc/s, and can be operated from a 110 V or 220 V d.c. mains supply, or from 24, 100, or 200 V d.c., or 230 V a.c. using a supplementary power supply unit.

The complete equipment, apart from the aerial system, including the goniometer, receiver, and power supply unit, is housed in a sheet metal case having approximate dimensions of 18 in. × 18 in. × 18 in., which is designed for bench mounting. The Lodestone IV, as distinct from the Lodestone III, has facilities incorporated for coupling to a gyro compass so that both true and relative bearings are shown simultaneously on a single scale by means of two separate cursors.

The goniometer, which is fitted within a protective magnetic screen, is mounted centrally and occupies a large part of the interior space of the case. The greater part of the front panel of the case slopes back at an angle of 30° from the vertical, and on this surface are mounted the three principal D.F. controls: namely, the *radio-goniometer knob*; the *tuning control*; and the *zero-sharpening control* (see Fig. 3.2). The large goniometer scale, which has a diameter of 10 in., is made of Perspex and is illuminated internally to facilitate precision and ease of reading. The goniometer pointer takes the

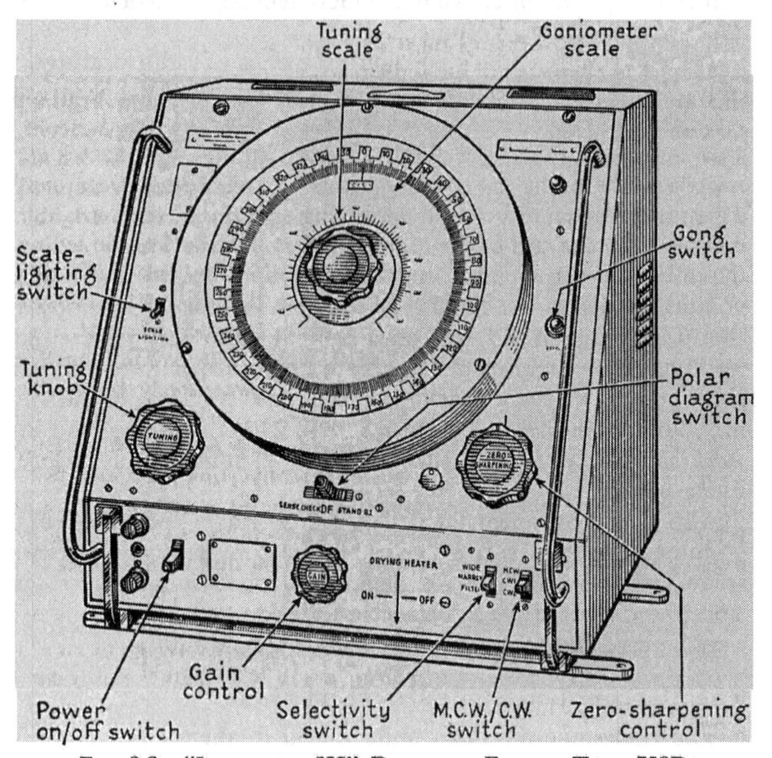

FIG. 3.2. "LODESTONE III" DIRECTION FINDER, TYPE 758D

form of a rigid Perspex disc carrying a blue cursor line. The tuning pointer is mounted on a sleeve concentric with the goniometer shaft, and is turned by means of a spring-loaded drive from the tuning condenser.

For ease of access, the goniometer and tuning scales are mounted outside the main panel, and are protected by a cowling which projects about 2 in. above the panel surface. The cowling is removable so that access to the scale-illuminating lamps is possible.

The tuning and zero-sharpening controls are located on the lower left- and right-hand corners of the sloping panel. Between these two controls is located the polar *diagram switch*, which is a four-position lever switch by means of which reception corresponding to any one of four polar diagrams can be selected. For normal D.F. reception, that is, two maxima and two minima, or figure-of-eight reception, the diagram switch is at right-angles to the panel surface at the position marked D.F. If the switch is clicked over to the right to the position marked STAND BI, reception is omnidirectional, in which case the polar diagram of reception is a circle. If the diagram switch is clicked over to the left of the D.F. position to the position marked SENSE CHECK 1, the reception is suitable for sensing, in which case the polar diagram of reception is a cardioid. The fourth position of the diagram switch is obtained by depressing the switch, against a spring, over to the left of the SENSE CHECK 1 position to a position marked SENSE CHECK 2, in which case the polar diagram of reception is again a cardioid, but reversed through 180° from the cardioid corresponding to the SENSE CHECK 1 position of the diagram switch.

On the Perspex scale bearing the visual degree markings where figures would normally be expected, there are transparent spaces alternating with opaque sections. Through the spaces there is visible a set of figures printed on a lower disc, and with the diagram switch in any of the first three positions, namely, D.F., STAND BI, or SENSE CHECK 1, these figures read from 0° at the top around to 360° in 10° steps. Movement of the diagram switch to the SENSE CHECK 2 position causes a rotation of the lower scale through 5°, thus causing a second set of figures to be brought into view, each being 180° different from the figure of the first set which occupied the position immediately before the diagram switch was moved to the SENSE CHECK 2 position. Thus the 0° space will now be filled with 180°, the 10° space with 190°, and so on.

The goniometer contains two search coils mounted at right-angles to one another. One is used for figure-of-eight reception and the other for sense determination. The second search coil eliminates the conventional necessity of swinging the bearing pointer through 90° when sensing. When the diagram switch is in either of the sense check positions, a vertical aerial signal is introduced which is in antiphase or in phase with the loop aerial signal, according to whether the diagram switch is in the SENSE CHECK 1 or SENSE CHECK 2 position for correct sense indication or for reciprocal sense indication respectively.

When the diagram switch is in the D.F. position for figure-of-eight reception, the vertical aerial is coupled to the search coil used for figure-of-eight reception, by way of a variable condenser, in order to provide for zero sharpening. For omnidirectional reception the vertical aerial signal alone is used.

Auxiliary controls necessitating less frequent use than the three principal controls discussed above are—

1. Gain control.
2. Selectivity switch.
3. Power switch.
4. Scale-lighting switch.
5. Gong switch.
6. Drying heater switch.
7. M.C.W./C.W. switch.

The front panel showing the relative positions of the D.F. controls is illustrated in Fig. 3.2.

The gain or volume control is located centrally on the lower vertical part of the front panel. The telephone jack socket and the power on/off switch are located to the left, and the selectivity switch and the M.C.W./C.W. switch are located to the right of the centre. The gong push-button switch and the scale-lighting switch are placed on the right and left sides of the sloping surface of the front panel.

The directional aerial with which the Lodestone direction finder is provided is of the Bellini-Tosi fixed loops type. The vertical aerial for sensing should be approximately 15 to 20 ft in length.

Calibration of Lodestone Direction Finder

In connexion with the calibration of a Lodestone direction finder, three important processes should receive careful attention. Firstly, before commencing calibration the D.F. site should be examined. A check should be made to ensure that the aerial loops are undamaged and that they are placed accurately on the centre line, one in the fore-and-aft line and the other in the athwartship line of the ship. All standing rigging in the vicinity of the aerial loops should be insulated efficiently so that no closed loops are formed. Any metal rails near the aerial loops should be examined to ensure that they do not make variable contact with their supporting stanchions. If this is not the case, such rails should be bonded or insulated. It is essential that the vertical sense aerial drops vertically from an insulated stay. Derricks and other movable metal gear should be in their sea-going positions.

During the calibration, the other two important processes are: the careful comparison between radio and visual bearings of the transmitter used for calibrating; and the balancing of the pick-up factors of the two loop aerial currents, by means of a calibration choke, in order to compensate for the quadrantal error caused by the metal structures of the ship.

After having checked the D.F. site as described above the receiver and its controls should be checked prior to commencing the calibration swing.

The initial calibration should never be hurried. Every effort should be made to ensure that the first calibration is accurate. Once the correct quadrantal error adjustment has been made, it should hold good for the life of the ship provided that no material change is made in the rigging or position of metal structures in the vicinity of the D.F. loops, and that the insulation or earthing of stays, stanchions, rails, etc. does not vary.

If the initial calibration produces indifferent or varying quadrantal error curves, the cause of the variation should relentlessly be pursued and rectified, otherwise the installation will always be unreliable. It is emphasized that to keep on recalibrating and rejecting earlier results is a waste of time because each new calibration will merely produce another curve of errors as unreliable as the previous one.

Calibration is essentially the process of finding the values of errors—other than variable errors—on all directions relative to the fore-and-aft line of the ship, and is carried out in the manner described on pages 58–9.

Because a ship's aerials may cause large errors in a radio bearing by re-radiating the incoming signals, there has been a long-standing rule that all aerials in the vicinity of the D.F. loops should be isolated whilst calibrating or using a direction finder for navigational purposes. It is impossible to lay down rules concerning the safe distance between the down-lead of a particular aerial and the D.F. loops, because the magnitude of the errors depend on a number of variable factors, such as the frequency to which the aerial is tuned. If the separation between the D.F. loops and the down-lead of any aerial exceeds 50 ft, then that aerial is unlikely to displace the D.F. bearing, and therefore it should not be necessary to isolate any aerial if its distance from the D.F. loop is more than 50 ft. It will be noted that it has been recommended in M.O.T.C.A. Notice No. M.402 that broadcast receivers should be attached to a single communal aerial, or to aerials which do not rise above the base of the loop aerial, or to aerials outside a radius of 50 ft.

The procedure for calibrating a Lodestone direction finder is as follows—

1. See that all nearby aerials other than the sense aerial are isolated, and that all movable gear is in its sea-going position.

2. See that no reception on or near the frequency used for calibration is taking place in the vicinity of the direction finder.

3. See that the calibration choke is disconnected.

4. Observe bearings of the transmitter ahead and astern and on each beam to ensure that the aerial loops and bearing pointer are correctly fixed.

5. Observe and record simultaneous radio and visual bearings of the transmitter every 5th degree around the ship.

6. From the record obtained plot a graph of errors and analyse the resulting curve. If the curve is a double sine curve with zeros at 0°, 90°, 180°, and 270°, the error is wholly quadrantal and may be eliminated by appropriate adjustment of the calibration choke. If, however, the curve includes a semicircular component, the cause must be traced and removed before the calibration can be considered to be complete and satisfactory.

If time is not available to carry out a complete calibration as described above, a quick calibration can be performed by adjusting the calibration choke until simultaneous radio and visual bearings of a transmitter, lying 45° on either bow, are equal. If the error is purely quadrantal then the calibration, as carried out by this method, will be correct. However, it must not be assumed that the direction finder is fully calibrated as the presence of any constant or semicircular error will not have been detected. The initial adjustment of the calibration choke should therefore be checked by observing simultaneous radio and visual bearings of the transmitter when it lies 45° on the quarter on the same side of the ship as the bow on which the first observations were made. If an error is then found, half the error should be removed by a second adjustment of the calibration choke. It will then be found that there is an error similar in magnitude but opposite in sign to that obtained on the first bearing. These two errors, it must now be assumed, are semicircular in character, and therefore radio bearings subsequently observed should be treated with reserve until such time as the cause of the error has been discovered and eliminated.

In order to observe a radio bearing by means of the Lodestone direction finder, the following procedure should be adopted—

1. Isolate all nearby aerials except those associated with the direction finder.

2. Switch on the power switch and allow about ½ min for the valves to warm up.

3. Switch on the scale-illuminating switch if required.

4. Set the diagram switch to the position marked STAND BI, for omnidirectional reception.

5. Set the selectivity switch to the position marked WIDE.

6. If A2 waves (continuous waves modulated at audible

frequency), are being transmitted, switch the M.C.W./C.W. switch to the upper position marked M.C.W. If A1 waves (C.W. unmodulated) are being transmitted, switch to the lower position marked C.W.1.

7. Set the zero-sharpening control index to zero.

8. Connect the headphones and tune the receiver to the frequency of the transmitter.

9. When the signal is heard, switch the selectivity switch to the position marked NARROW and retune carefully. Should the signal be weak or masked by interference, further advantage will be gained by switching the selectivity switch to the position marked FILTER, and retuning to give optimum output. In certain cases of unmodulated C.W. reception, where an interfering signal is fairly close in frequency to the desired one, the beat note produced can be displaced from the desired note (so resulting in less interference) by switching the M.C.W./C.W. switch to the position marked C.W.2, at which the local-oscillator frequency is 2 kc/s different from its frequency when the switch is in the C.W.1 position.

10. When the receiver is correctly tuned, set the diagram switch to the position marked D.F. The goniometer cursor should then be swung to ascertain the reciprocal readings on the goniometer scale at which minimum signals are heard. Under normal conditions each minimum position extends over several degrees of arc. To determine the exact centre of the arc the goniometer cursor should be swung to and fro through a signal minimum while simultaneously adjusting the zero-sharpening control until the arc of minimum is reduced to an absolute zero. To determine which of the two reciprocal bearings is the bearing of the transmitter, that is, to sense, the diagram switch is set to the positions marked SENSE CHECK 1 and SENSE CHECK 2 in turn, with the goniometer cursor set to one of the positions for minimum reception. Only one of the sense check positions will result in a minimum, and with the diagram switch in this particular position the correct bearing of the transmitter is shown on the goniometer scale.

11. At the instant of observing the radio bearing press the gong switch so that the officer of the watch may observe and record the compass heading of the ship at the time at which the radio bearing was observed.

12. If the D.F. installation has been properly calibrated, the observed radio bearing relative to the ship's head, applied to the true heading of the ship at the time of the observation, will yield the great-circle bearing of the transmitter from the ship. The half-convergency correction, if any, applied to this will give the

rhumb-line bearing which determines the direction of the radio position line to be plotted as a straight line on a Mercator chart.

The Marconi Direction Finder, Type 579

The instrument used by the Ministry of Transport and Civil Aviation examiners for the practical examination of candidates for the First Mate (Foreign-going), and Master (Home Trade examinations), on radio direction finding is the obsolescent Marconi direction finder, Type 579.

The Marconi direction finder, Type 579, may be tuned to wavelengths within the range 550–1,150 metres. The aerial system is of the Bellini-Tosi type. The aerial loops are connected to the crossed coils of a radio goniometer unit which is provided with a double-ended bearing pointer and a sense pointer at right-angles to it. One end of the bearing pointer rides within a fixed scale graduated from 0° at the top clockwise to 360°. The other end rides within a similar, but movable, graduated scale. The first-mentioned end of the pointer is labelled REL BEARING and is used to determine the bearing of a transmitter relative to the fore-and-aft line of the observer's ship. The other end is labelled TRUE BEARING and may be used to determine the true bearing of a transmitter, after having first set the movable scale so that the ship's true heading coincides with the zero mark on the fixed scale.

A braided flex connects the search coil and goniometer coils to the receiver unit, which occupies the lower part of a metal case, in the upper part of which is fitted: the search-coil tuning circuit; a sense-finding unit; a zero-sharpening circuit; and an oscillator circuit for heterodyning.

The Bellini-Tosi aerial which is used in the instrument should be fitted as high in the ship as practicable, and well clear of standing rigging and large metal masses. The sense aerial should have a length of about 15 ft and should be fitted vertically, usually by means of a halyard rove through a block on the triatic or jumper stay.

After setting up the direction finder, the vertical aerial circuit is adjusted such that, when the receiver is tuned to the frequency of a transmitter, the induced current in the sense aerial is equal to the maximum current induced in the loop aerial circuit. The vertical aerial, as well as providing for sensing, also serves as a means for neutralizing any effect which results in a distortion of the figure-of-eight reception so causing a wide minimum. The zero-sharpening property of the vertical aerial is brought into effect by use of the control knob marked D.F. BALANCE.

The controls fitted to the Type 579 direction finder are—

1. Main receiver tuning knob and scale.
2. Search-coil tuning knob.
3. Heterodyne wavemeter.
4. Power on/off switch.
5. D.F. balance.
6. Sockets for headphones.
7. D.F./sense switch.
8. Gain control.

The goniometer unit and that which houses the receiver and other circuits, together with the direction finder controls, are illustrated in Fig. 3.3.

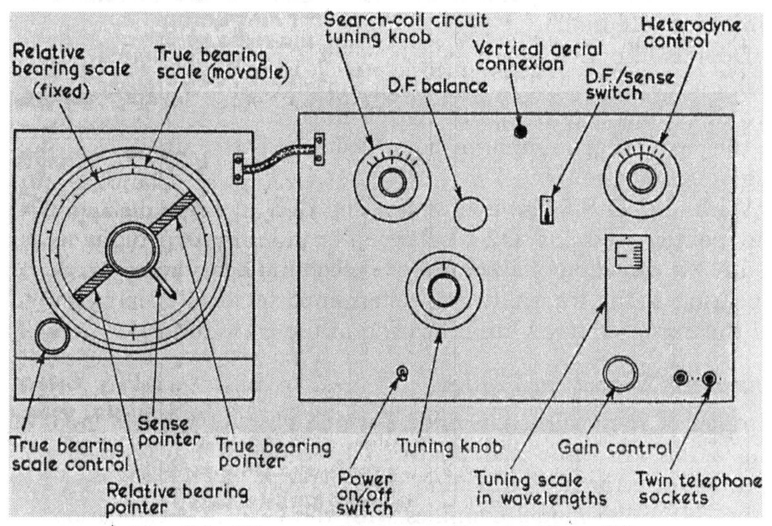

FIG. 3.3. MARCONI DIRECTION FINDER, TYPE 579

In order to observe a radio bearing of a transmitter by means of the Type 579 direction finder, the following procedure should be adopted—

1. Plug in and fit the headphones.
2. Switch on the power switch.
3. By means of the receiver tuning knob set the index on the wavelength tuning scale (visible in an illuminated window) to the wavelength of the transmitter.
4. If the transmitter is working with A1 waves (C.W. unmodulated), set the index on the heterodyne wavemeter to the station's wavelength.

5. Adjust the search-coil tuning dial to the wavelength of the transmitter.

6. Swing the bearing pointer of the goniometer, at the same time increasing the gain or volume, until an audible signal is obtained.

7. Retune if necessary for best reception.

8. Select the pitch of the audible signal by adjusting the heterodyne wavemeter. (If A2 waves—C.W. modulated at audible frequency—are being transmitted, as is the case with radio beacons, the heterodyne wavemeter ought to be set away from the wavelength of the transmitter in order to avoid possible interference.)

9. Swing the bearing pointer and determine the two directions for minimum reception.

10. Depress the D.F./sense switch, which is spring-loaded, to the position marked SENSE, and bring the sense pointer to each of the two directions determined in step 9, in turn. The direction in which the weaker, or minimum, signal is heard is the approximate bearing of the transmitter.

11. With the D.F./sense switch to the D.F. position, use the bearing pointer and the D.F. balance control simultaneously to determine the direction of absolute zero reception near the approximate bearing of the transmitter as determined in step 10, in order to find the exact observed radio bearing of the transmitter.

The Siemens Direction Finder

Examples of radio direction finders which employ a single loop rotating aerial are those manufactured by Siemens Edison Swan Limited. Of the range of direction finders produced by this company the R19/SB85 model complies with the G.P.O. specification for a radio direction finder for compulsory fitting to ships.

The R19/SB85 equipment is suitable for direction finding with A1 (C.W. unmodulated) or A2 (C.W. modulated for audible frequency) transmissions, a beat frequency oscillator being provided. The frequency range is 225–525 kc/s, and the equipment can be operated from 110 V or 200 V d.c., or from 250 V a.c. In the case of d.c. mains a rotary converter is provided to suit the mains voltage.

The main receiver, sense finder, and oscillator circuit are housed in a metal case having dimensions of 21½ in. width, 16 in. depth, and 14 in. in height. The power unit and the rotary converter (if required) are accommodated in separate cases and these two units are fitted in any convenient space in the chartroom or radio office.

The screened frame, within which is fitted the loop aerial, is direct driven from a handwheel, and has a diameter of 3 ft. The handwheel, which is designed for mounting on a bench, carries a scale of degrees from 0° to 360°. Concentric with the scale is a device by means of which quadrantal error is automatically corrected. The device takes the form of a cam, the shape of which is determined from the curve of errors produced from the results of the calibration swing which is made when the equipment is fitted.

The "Adsum 1" Automatic Visual Direction Finder

The "Adsum 1," a product of the International Marine Radio Company Limited, is an automatic direction finder of ultra-modern design, giving a direct visual bearing indication on a cathode-ray tube.

A cathode-ray tube (C.R.T.) is a device in which a stream of electrons is emitted from a heated filament, or *cathode*, and focused on to a screen, forming the face of the tube, on which a fluorescent material produces a luminous spot at the point where the stream of electrons strikes it. The electron stream, or *cathode ray* as it is called, can be deflected by electrostatic or electromagnetic forces, such that it can strike and produce a luminous spot at any point on the screen.

The deflexion system in the C.R.T. incorporated in the "Adsum 1" comprises two pairs of parallel flat plates which are mounted within the tube; one pair is set vertically and the other horizontally. The vertical pair causes the electron stream to be bent horizontally, and the horizontal pair causes the stream to be bent vertically. Since a positive voltage attracts electrons and a negative voltage repels them, the electron stream in the C.R.T. can be moved off the axis of the tube by voltages applied to the deflexion plates. The amount of deflexion depends upon the magnitude of the voltage applied to the plates. A varying voltage applied to either pair of deflexion plates would cause the electron stream to sweep across the tube face, and consequently produce, or *paint*, a luminous line or *trace*.

The "Adsum 1" direction finder is equipped with an aerial assembly consisting of a pair of fixed crossed loops, one in the fore-and-aft line and the other in the athwartship line of the ship. The vertical pair of deflexion plates is associated with the fore-and-aft loop, and the horizontal pair with the athwartship loop. Each of the loop aerials is connected to its respective deflexion plates by way of a *twin-path receiver*. This may be regarded as two receivers, one of which deals with the signal voltage induced in the fore-and-aft loop, and the other with the signal voltage induced in the athwartships loop. The two paths are interconnected to use a common local oscillator; common phase and amplitude networks; and a common audio output to a loudspeaker for monitoring purposes.

The twin-path receiver, as well as being fed with the signal voltages induced in the loop aerials, also receives the output from an omnidirectional sense/balance aerial which is in the form of a 9 ft long rod fitted in the vertical axis of the 4 ft diameter D.F. aerial loops.

The frequency of the signal voltage induced in each of the two loops is reduced, and the alternating voltage is applied to the respective deflexion plates of the C.R.T. It follows that, because

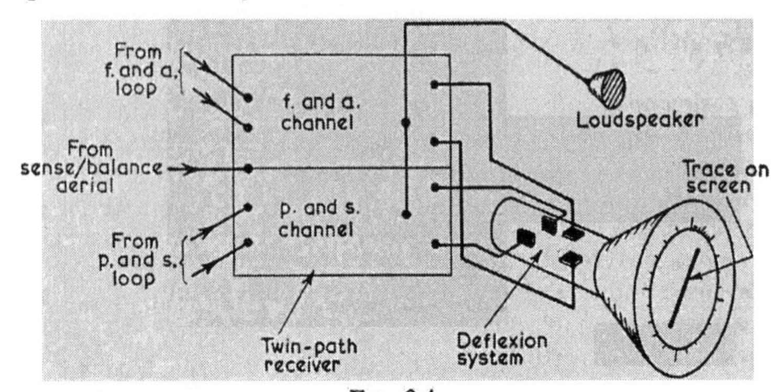

Fig. 3.4

the plates of the deflexion system are all equal in size, and symmetrically placed in relation to the axis of the C.R.T., the electric forces on the electron stream, acting between opposite plates, will be proportional to the e.m.f.s induced in the aerial loops. The direction of the luminous trace on the screen will thus be representative of the resultant of the induced e.m.f. in both loops, just as the direction of the resultant magnetic field, determined by a search coil in a goniometer, is representative of the resultant e.m.f. induced in the loop aerial.

The bearing of a transmitter is determined on the "Adsum 1" by the direction of the luminous trace, and is indicated on a 0°—360° scale which is fitted around the circumference of the screen.

The general arrangement of the "Adsum 1" direction finder is illustrated in Fig. 3.4.

The two loop voltages must be equal in amplitude and phase to ensure correct bearing indication. This is achieved by "line" (phase) and "angle" (amplitude) correction controls.

Determination of sense is achieved by means of a "reversed" goniometer, the search coil of which is connected to the sense aerial when the main switch is placed in either of two sense positions. In these positions, the fixed coils of the goniometer are connected to the loop aerials, thus superimposing the sense aerial signal voltages on the loop aerial signals in suitable relative amplitude and phase relationships. Correct sense is indicated by a substantial shortening, and incorrect sense by a lengthening, of the luminous trace.

The whole assembly, including receiver, power supply unit, error compensator unit, and cathode-ray indicator, is housed in a metal

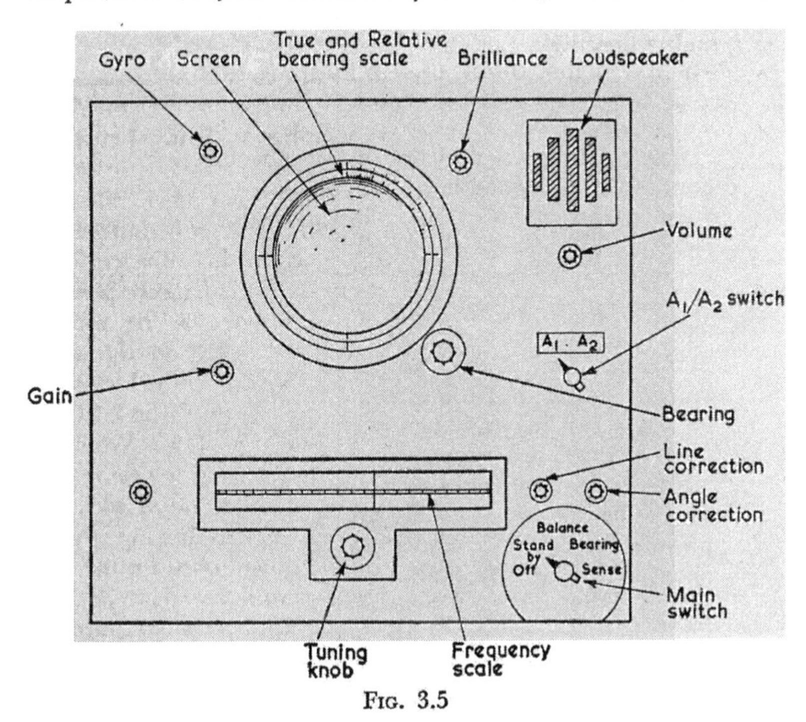

FIG. 3.5

cabinet. The instrument has a frequency range of 240–535 kc/s, and may be used for receiving A1 or A2 transmissions. The receiver includes an extensive test system for checking valves and voltages.

The outstanding features of the "Adsum 1" direction finder are—

1. Automatic visual indication of true and relative bearings on a C.R.T.

2. No headphones, or careful zero determination, necessary as with aural null direction finders.

3. Once tuned to a radio beacon the bearing can be observed continuously whilst the beacon is transmitting.

4. No difficulties in observing bearings in the presence of interfering transmissions, as the bearings are based on maximum and not minimum signal strength.

5. Easy determination of bearings when a radio beacon transmission is weak. Ambiguity due to difficulty of obtaining a clear zero is obviated.

6. Instantaneous assessment of the reliability of a bearing by visual observation only. An important feature in the case of night effect.

7. Simultaneous monitoring of audible signal on loudspeaker whilst taking bearings.

8. Easy observation irrespective of local external noise such as a ship's syren.

The "Adsum 1" direction finder, together with its operational controls, is illustrated in Fig. 3.5.

In order to observe a radio bearing with an "Adsum 1" the main switch is set to the stand by position and the receiver is tuned to the frequency of the transmitter. The volume control is adjusted for an audible signal. The main switch is then set to the balance position. This connects the two radio-frequency outputs of the receiver in parallel and introduces equal signal voltage from the sense/balance aerial to both inputs. Correct balance is achieved by adjusting the line and angle correction controls until the luminous trace on the C.R.T. screen lies along the line 045°–225°. When this has been done the main switch is set to the bearing position. This causes the line and angle controls to be declutched mechanically to prevent inadvertent alteration of the balance during further operation. The main tuning control is also declutched for the same reason. After the main switch has been set to the bearing position, the gain and brilliance controls are adjusted to produce a satisfactory trace, and the bearing, true or relative, is then read off.

Part 4

Consol

CONSOL is a long-range aid to navigation used principally by air navigators. It may, however, be of value to mariners for the determination of approximate position lines when in the coverage area of the Consol system. Fixing or determining a position line by Consol normally involves the use of the ship's radio direction finder, but any suitable radio receiver, capable of receiving signals transmitted on the frequency used by a Consol station, may be used.

At the present time there are five Consol stations in operation, and the service details of these are to be found in *The Admiralty List of Radio Signals*, Volume 5, in which will also be found a brief description of the Consol system of navigation, together with a chartlet showing the boundaries of the coverage provided by each of the stations and tables to facilitate the use of Consol. The five Consol stations are located in north-west Europe. In 1960 a long-range navigational system similar to Consol was brought into use in the U.S.A. This system is known as *Consolan,* and two Consolan stations—one at Nantucket and the other at San Francisco—were established during this year.

At a Consol station, medium-frequency transmissions are emitted from a directional aerial system consisting of three aerials equidistantly spaced and sited in the same straight line. The distance between the outer aerials is about six times the wavelength of the radiated energy, the latter being of the order of 1,000 metres.

The radio signals from a Consol station, on being received, produce after heterodyning—in order to render the signal audible—either shorts (dots) or longs (dashes). The transmissions which produce the dots and dashes have a duration of either 30 or 60 sec. At the beginning of each transmission cycle the radiation pattern around a Consol station comprises a number of sectors centred at the station as illustrated in Fig. 4.1 in which A represents the Consol station.

It will be noticed from Fig. 4.1 that the angular width of the sectors is variable, being least in a direction at right-angles to the alignment of the aerials at the station. The sectors, it will also be noticed, are labelled with a dot (·) or a dash (–), indicating that dots (or dashes) are heard at the beginning of each transmission cycle in alternate sectors.

Consol, as will be described later, is a hyperbolic system of

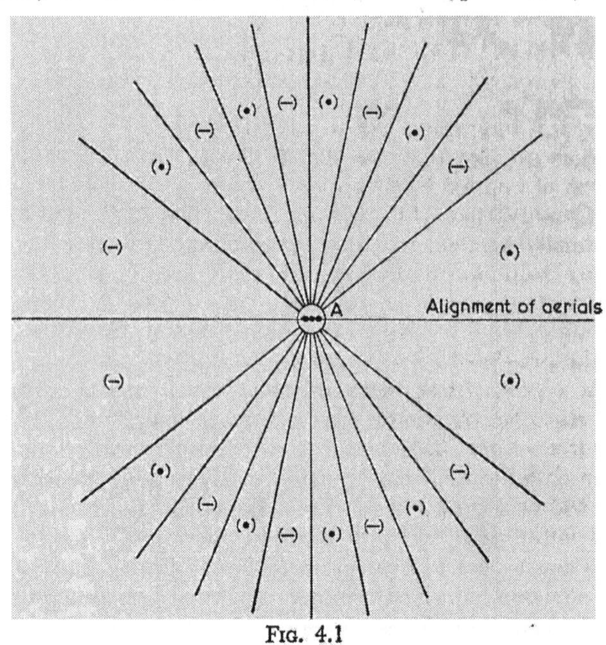

FIG. 4.1

navigation, the limits of the dot and dash sectors forming a series of hyperbolae whose foci are located at the outer aerials of the station.

Properties of Hyperbolae

A hyperbola, it will be remembered, is one of a pair of curves produced at the section of a pair of circular cones by a plane surface which is parallel to the axes of the cones, as illustrated in Fig. 4.2.

Mathematicians define a hyperbola (or any of the other conics, viz., the parabola and the ellipse) in terms of a fixed point and a fixed straight line known as the focus and directrix of the conic.

Referring to Fig. 4.3, if AB is the directrix and F the focus of a conic, then if P traces a path represented by the dotted line in the figure, such a path is the conic if the ratio FP/PQ is constant.

The ratio FP/PQ is known as the eccentricity of the conic. If the eccentricity is less than unity, the curve is an ellipse; if it is exactly unity, the curve is a parabola; if it is greater than unity, the curve is a hyperbola.

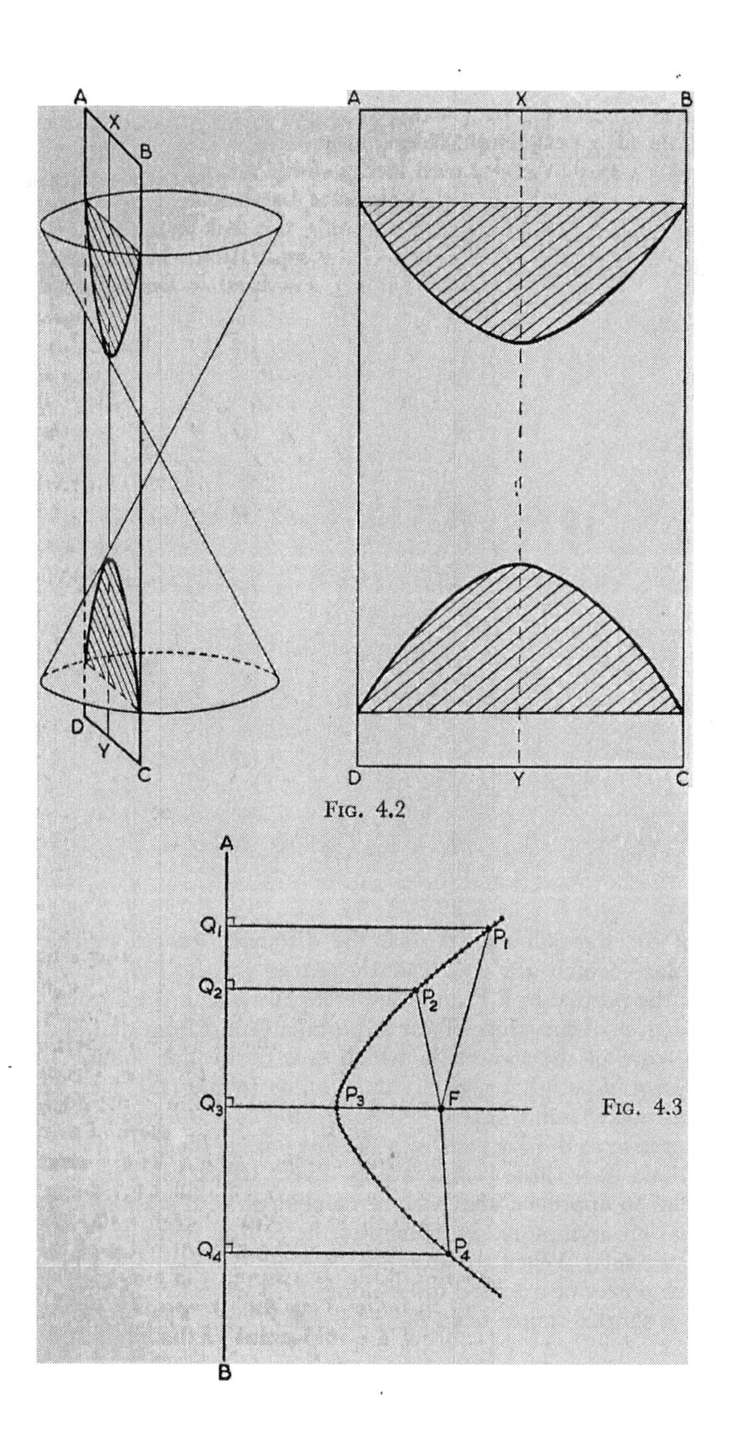

FIG. 4.2

FIG. 4.3

A set of hyperbolae having common foci forms a family of confocal hyperbolae as depicted in Fig. 4.4.

Fig. 4.4 reveals that the distance between any two adjacent hyperbolae is least on the line joining the foci F_1 and F_2. As the distance from this line increases, the separation also increases. It will also be noticed that, at any given distance from the central

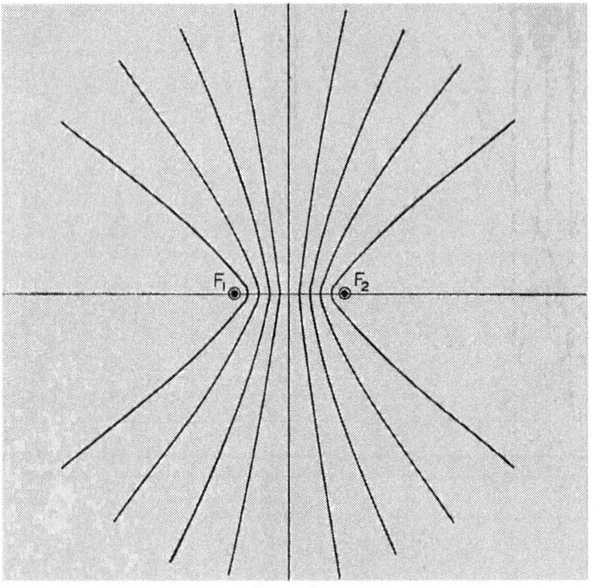

FIG. 4.4

point of the line joining the foci, the distance between adjacent hyperbolae—which are equidistantly spaced on the line F_1F_2—is least on the normal to F_1F_2, and increases away from the normal.

The degree of curvature of any hyperbola (apart from that of the limiting case of the hyperbola which crosses the line joining the foci at its mid-point) is greatest on the line joining the foci, and decreases away from this line. Although the curvature of a hyperbola decreases as the distance from the line joining the foci increases, a hyperbola never loses its curvature. Every hyperbola does, however, tend to approach and become coincident with a straight line known as an asymptote, on which lies the central point of the line joining the foci. When the distance between the foci is short, the hyperbolae may be assumed to lie along their respective asymptotes at only a short distance from the mid-point of the line joining the foci.

For the purposes of understanding the principles of hyperbolic navigation, a hyperbola may be defined in terms of one of its important geometrical properties. Thus, a hyperbola is a curve such that the difference in distances from two fixed points to any point on the curve is constant. The two fixed points referred to in the definition, and illustrated at A and B in Fig. 4.5, are the foci of the family of confocal hyperbolae to which the curve XY belongs. The curve XY is a hyperbola if (AP–BP) is constant for all positions of P.

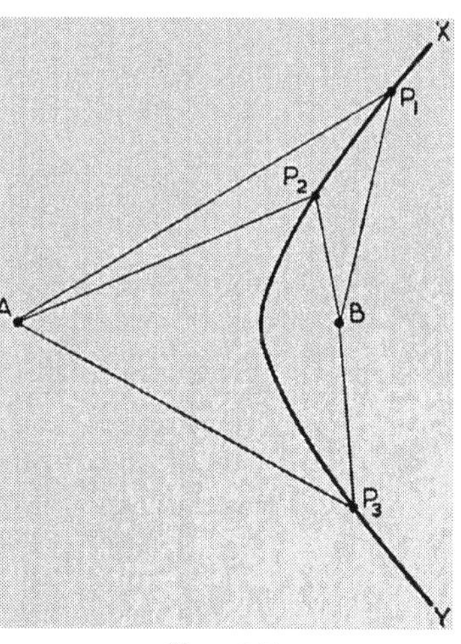

Fig. 4.5

Principles of Hyperbolic System of Navigation

The three aerials of a Consol station, during each transmission cycle, radiate continuous unmodulated waves. The amplitude of the waves radiated from the centre aerial is greater than that of the waves radiated from the outer aerials. In practice the amplitude of the waves from the centre aerial is four times the amplitude of the waves from the outer aerials. During the transmission cycle the phase of the waves from the outer aerials varies with respect to those from the centre aerial. This has the effect (which will be explained later) of causing the dot and dash pattern to rotate around the Consol station.

At the commencement of each transmission cycle, the phase of the current fed to the outer aerials differs by 90° from the current fed to the centre aerial. That fed to one of the outer aerials lags behind the current fed to the centre aerial, and that fed to the other outer aerial leads on the current fed to the centre aerial. Treating the currents fed to the three aerials as vector quantities, they may be represented graphically as in Fig. 4.6, in which OX represents the current fed to the centre aerial; OY the leading current fed to one of the outer aerials, and OZ the lagging current fed to the other outer aerial.

It will be noticed that at the beginning of each transmission cycle, as illustrated in Fig. 4.6, the phase difference between the currents fed to the outer aerials is 180°.

The strength of the signal received at any point in the coverage area of a Consol station will be dependent upon the distance of the point from the station, and also upon the relative phases of the signals received from each of the three transmitting aerials. If, at the beginning of the transmission cycle, each of the three component signals arrived in the same phase relation in which they left the transmitting station, the resultant signal would be the same as the signal received from the centre aerial only, since those from the outer aerials, having a phase difference of 180°, would cancel one another. This condition would apply at all receiving stations

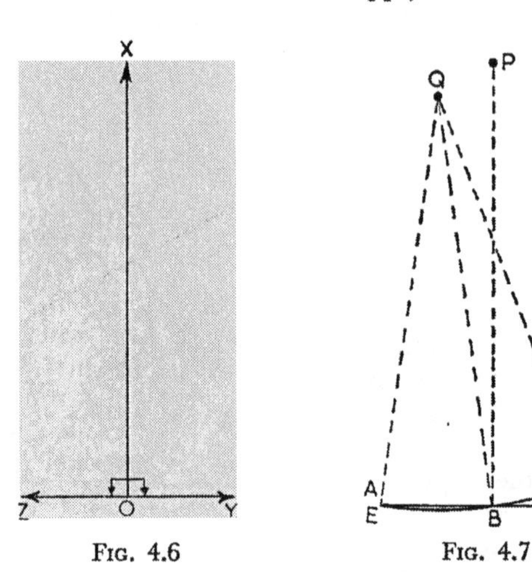

Fig. 4.6 Fig. 4.7

located on the perpendicular bisector of the line joining the three aerials.

Referring to Fig. 4.7, imagine a ship at P, whose receiver is tuned to the Consol station, the aerials of which are located at A, B, and C, to be situated on the perpendicular bisector of the line joining A, B, and C. The radio energy from A and C, having the same distance to travel to P, will, on arriving at P, have no effect on the strength of the signal received, because of the phase difference of 180°. This assumes that the rate of travel of the radio energy from A to P is the same as that from C to P. Now suppose a ship to be located at position Q in Fig. 4.7. The energy arriving at Q from each of the three aerials will have different distances to travel. The distance from C to Q is greater than the distance from B to Q, and the distance from A to Q is less. Now the differences between these distances are small if AC is small compared with BQ, and will not therefore materially affect the signal strength due to the combined waves from A, B, and C. But the differences in distances are large compared with the wavelength of the radio energy transmitted, and the relative phases of the signals arriving at Q will therefore be affected. The signal from aerial C will take longer to reach the ship than that from B by an amount proportional to CD. The waves on arrival at Q will therefore be retarded in phase by an amount proportionately to CD. Similarly the waves from A will be advanced in phase by an amount proportional to AE. If either CD or AE

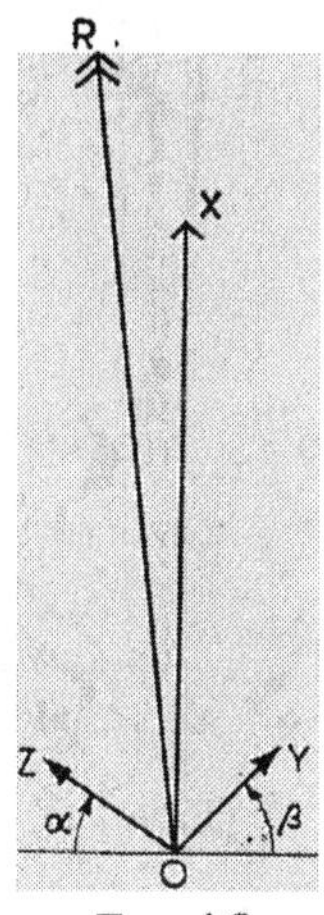

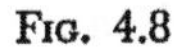
FIG. 4.8

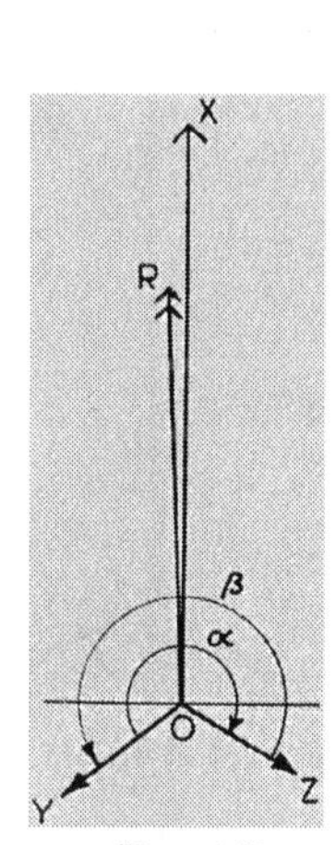

FIG. 4.9

happens to be equal in distance to one wavelength, the advance or retardation in phase would be 360°. If CD or AE were n wavelengths the advance or retardation in phase would likewise be 360°.

If the waves from A on arrival at Q are advanced in phase by $\alpha°$, and those from C are retarded in phase by $\beta°$, the resultant amplitude, represented by OR in Fig. 4.8, will be greater than that of the signal from B alone.

At certain other points in the coverage area, the vector sum of the signals from A, B, and C will be less than the signal from the centre aerial only, as illustrated in Fig. 4.9.

KEYING

The dot and dash pattern of the Consol transmission is produced by *keying*, a process in which the phase of the current fed to each of the outer aerials is changed suddenly by 180°.

Referring to Fig. 4.8, if the transmissions from the outer aerials are suddenly changed in phase by 180°, there will be a sudden reduction in the signal strength, as the resultant of the signals from the three aerials after keying will now be less than the signal received from the centre aerial only, as illustrated in Fig. 4.10. Referring to Fig. 4.9, if the transmissions to the outer aerials are suddenly changed in phase by 180°, there will be a sudden increase

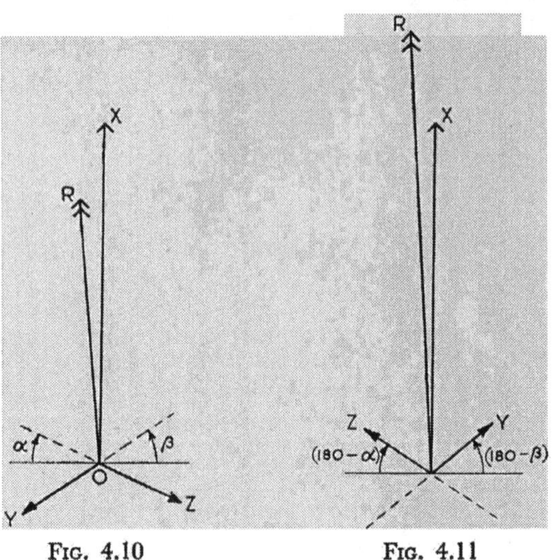

FIG. 4.10 FIG. 4.11

in the signal strength, as the resultant will now be greater than the signal from the centre aerial only, as illustrated in Fig. 4.11.

During each transmission cycle, the aerials are keyed at intervals of $\frac{1}{8}$ sec and $\frac{3}{8}$ sec when the transmission period is 30 sec, and $\frac{1}{4}$ sec and $\frac{3}{4}$ sec when the period is 60 sec. At point Q (refer to Fig. 4.7), therefore, if the transmission period is 30 sec, the stronger signal is received for $\frac{3}{8}$ sec and the weaker signal for $\frac{1}{8}$ sec, and a series of dashes, having a period of $\frac{1}{2}$ sec, will be heard. At certain other points, where the resultant signal before keying was less than the resultant signal after keying, the stronger signal will be received for $\frac{1}{8}$ sec, and the weaker signal for $\frac{3}{8}$ sec, during each successive $\frac{1}{2}$ sec period, and a series of dots, as a consequence, will be heard.

Referring again to Fig. 4.7, if the ship moves from the point Q, the distances AE and CD will change, and consequently the angles α and β will also change. These angles will, at certain points in the coverage area, be such that the vectors OY and OZ will be 180°

apart, and, when this is so, the signals from the outer aerials will cancel one another, as illustrated in Fig. 4.12. After keying, the same will be true as depicted in Fig. 4.13.

At positions where the vectors OY and OZ represent signals which have a phase difference of 180°, keying has no effect on the strength of the signal received, and a continuous sound signal will be heard at such positions. At all other positions in the coverage

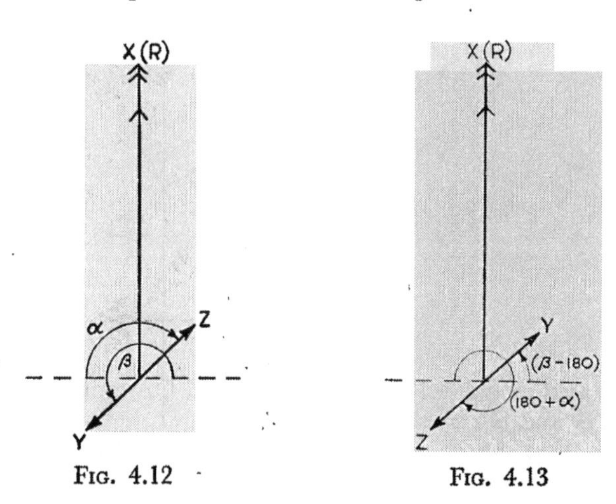

FIG. 4.12

FIG. 4.13

area, either dots or dashes are heard at the beginning of each transmission cycle.

EQUISIGNAL LINES

The continuous sound signal referred to is known as an *equisignal*, and lines joining points where the equisignal may be heard are known as *equisignal lines*.

The difference in time for the signals to travel from the outer aerials to any point on an equisignal line is constant, as the phase difference between the waves on arrival is, in all cases, 180°. The equisignal lines form, therefore, a family of confocal hyperbolae, with the foci located at the outer aerials of the Consol station. Since the distance between the aerials is small, the hyperbolae may be assumed, for practical purposes, to lie on asymptotes at ranges greater than about twelve times the aerial spacing. The equisignal lines may thus be regarded as arcs of great circles passing through the position of the centre aerial.

At distances of more than about twenty miles from the Consol station, the directions of the ship from each of the three aerials may be considered as being parallel, as illustrated in Fig. 4.14. In these cases, the distances AE and CD are equal.

In Fig. 4.14, it will be observed that the difference in the distances from C and B to the ship is equal to the difference in the distances from A and B to the ship. That is to say AE = CD. The phase lead of the signal from A on the signal from B will therefore be equal to the phase lag of the signal from C behind the signal from B. The vector diagram of the signals from the three aerials is

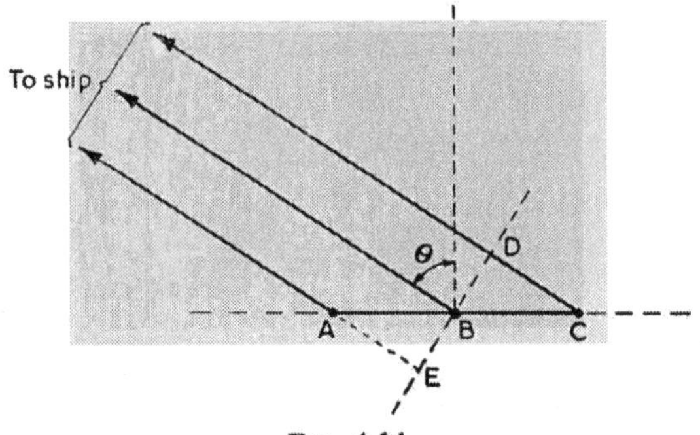

Fig. 4.14

illustrated in Fig. 4.15, in which OY and OZ represent the signals from A and C respectively.

The ship lies on an equisignal line when the vectors OY and OZ cancel one another, and this will occur when $\alpha = \beta = 0°$, 180°, 360° . . . $n180°$. When $\alpha = \beta = 180°$, the distances AE and CD (refer to Fig. 4.14) are each equal to half the wavelength of the radiated energy.

The ship lies on an equisignal line when AE or CD is $n \,.\, \lambda/2$, where n is a whole number and λ is the wavelength of the transmitted radio energy.

From Fig. 4.14

$$\text{AE or CD} = \text{AB} \sin \theta$$

where θ is the angle between the direction of the ship from the centre aerial and the normal to the line joining the aerials.

If the distance between the outer aerials is d, then

$$\text{AE or CD} = d/2 \,.\, \sin \theta$$

Thus

$$n \,.\, \lambda/2 = d/2 \,.\, \sin \theta$$

from which

$$\sin \theta = n \,.\, \lambda/d$$

θ in this case being the angle between the normal to the line of aerials and the direction of an equisignal line from the centre aerial. From this relationship, the directions of the equisignal lines may be determined.

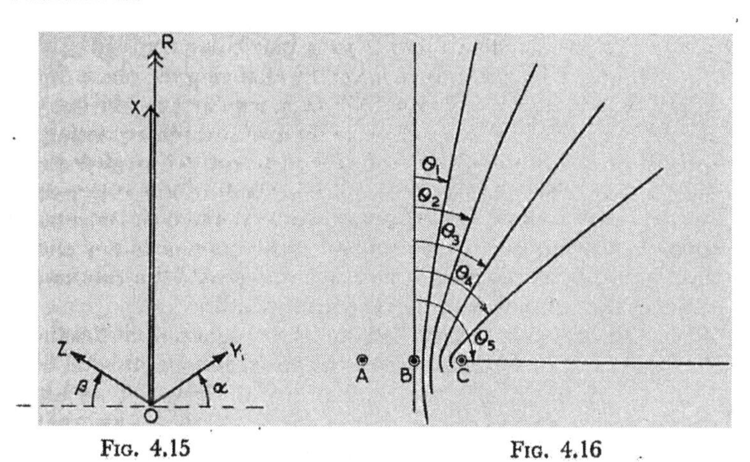

Fig. 4.15

Fig. 4.16

If $\lambda = 1{,}000$ metres, and $d = 5{,}000$ metres, the directions of the equisignal lines are found as follows—

$$n = 0,\ \sin\theta_{(n=0)} = 0 \times \frac{1{,}000}{5{,}000} = 0 \cdot \quad \therefore\ \theta_{(n=0)} = 0^\circ$$

$$n = 1,\ \sin\theta_{(n=1)} = 1 \times \frac{1{,}000}{5{,}000} = 0{\cdot}2 \quad \therefore\ \theta_{(n=1)} = 11\tfrac{1}{2}^\circ$$

$$n = 2,\ \sin\theta_{(n=2)} = 2 \times \frac{1{,}000}{5{,}000} = 0{\cdot}4 \quad \therefore\ \theta_{(n=2)} = 23\tfrac{1}{2}^\circ$$

$$n = 3,\ \sin\theta_{(n=3)} = 3 \times \frac{1{,}000}{5{,}000} = 0{\cdot}6 \quad \therefore\ \theta_{(n=3)} = 37^\circ$$

$$n = 4,\ \sin\theta_{(n=4)} = 4 \times \frac{1{,}000}{5{,}000} = 0{\cdot}8 \quad \therefore\ \theta_{(n=4)} = 53^\circ$$

$$n = 5,\ \sin\theta_{(n=5)} = 5 \times \frac{1{,}000}{5{,}000} = 1{\cdot}0 \quad \therefore\ \theta_{(n=5)} = 90^\circ$$

The results of this example are illustrated in Fig. 4.16.

It will be noticed from Fig. 4.16 that the angles between successive equisignal lines are not uniform, being least when n is small and increasing as n increases.

The pattern of the equisignal lines which delimit the alternate dot and dash sectors is made to rotate, thus enabling an observer to determine a position line relative to a particular equisignal line. The rotation of the pattern is achieved by changing the phase of the current to the outer aerials by 180° at a regular rate during the transmission cycle. The current to one of the outer aerials is retarded at the same uniform rate as the current to the other outer aerial is advanced. At the end of each transmission cycle, each equisignal line occupies the position which at the commencement of the cycle was occupied by the equisignal line immediately ahead of it. On one side of the line joining the aerials the rotation is clockwise, and on the other side it is anticlockwise.

On an equisignal line the resultants of the signals from the three aerials, before and after keying, are equal. Near an equisignal line, the resultants before and after keying are nearly equal, and it is therefore difficult to distinguish, by ear, the change of loudness of the signal due to keying. This has the effect of broadening the equisignal, the narrowness or sharpness of which depends upon the ratio between the amplitudes of the waves from the centre and outer aerials, which, as stated before, is 4:1.

Determination of Consol Counts

During the transmission cycle, keying is performed 60 times; consequently 60 characters (dots or dashes) should be heard during each keying cycle. Because of the width of the equisignal line, some characters merge together to form a prolonged signal, and therefore, in practice, less than 60 characters are counted. The number of characters counted depends upon the ability of the observer to distinguish changes in the loudness of the resultant signal before and after keying, when the equisignal line approaches or recedes from his position. Because the period of the dot signals is the same as that of the dash signals, the equisignal is comprised of an equal number of dots and dashes.

When determining a position line by means of Consol, the radio receiver—which may be a radio direction finder receiver—is tuned to the frequency of the Consol station. Following a continuous signal and the call sign of the station, keying commences, and the observer counts the number of dots and dashes (which are separated by the equisignal) during the keying cycle. The total number of dots and dashes in the observed count, being less than 60, must be corrected by subtracting the total from 60, and adding half the remainder to the observed count of the first character heard. For example, if the observed count is 16 dots, equisignal, 40 dashes, the corrected count is

$$16 + \left(\frac{60 - 56}{2}\right) = 18 \text{ dots}$$

Some observers are able to distinguish dots easier than they can distinguish dashes. In this event their observed counts will contain more dots and fewer dashes than would be the case if dots and dashes are distinguished with equal facility. Much practice at making Consol counts is therefore necessary, so that the possibility of error due to this cause is reduced or eliminated. It is prudent always to make several counts and to strike an average before consulting Consol tables or charts.

The arc swept out by an equisignal line during the keying period (either ½ sec or 1 sec) is not uniform throughout the cycle, for the same reason as the angle at the centre aerial between adjacent equisignal lines is not uniform. Bearing in mind, however, that as many as 60 characters are transmitted during the time taken for the equisignal line to swing through an angle the value of which is of the order of 10° or 15°, it may be assumed that the equisignal line does in fact swing at a uniform angular rate. If therefore the count is determined to the nearest one character, it follows that the bearing of the ship from the station can be found with a degree of accuracy of the order of ¼° or thereabouts, if it is known in which sector the ship is located. There will be ambiguity unless the bearing of the Consol station is known to within about 10°. If the ship's D.R. position cannot be relied upon to resolve any ambiguity, the ship's radio direction finder may be used to determine the approximate bearing of the Consol station, the continuous note and the station's call sign, which are transmitted before each keying cycle, providing the means for determining a radio bearing of the station.

If, when making a Consol count, dots are heard before the equisignal, the observer is in a dot sector. If dashes are heard first the ship is in a dash sector. If the corrected count is 60 dots or 60 dashes, the observer is located on an equisignal line. If the corrected count is 30 dots or 30 dashes, he is located midway between adjacent equisignal lines—ones bounding a dot sector if the corrected count is 30 dots, and ones bounding a dash sector if the count is 30 dashes. In all cases, after having made and connected a Consol observation, the great-circle bearing of the station may be found by means of Consol tables, which are published in *The Admiralty List of Radio*

Signals, Volume 5. It must be remembered that the bearings given in Consol tables are great-circle bearings, and, before laying down a Consol position line on a Mercator chart, the necessary half-convergency correction must be applied to the great-circle bearing to obtain the rhumb-line bearing.

The most satisfactory method of using the Consol system of navigation is by means of a Consol chart. This is an ordinary navigational chart overprinted with a lattice of lines marked with count numbers and characters corresponding to corrected Consol counts. The Consol position lines—which form the lattice of the chart—are printed in different colours for the different stations whose coverage areas lie within the charted area.

When using a radio direction finder for the purpose of obtaining a Consol count, the loop aerial or the goniometer search coil should be set to the position for maximum reception. It may be necessary, in order to suppress interference, to rotate the loop or search coil away from this position, but, if this is necessary, it should not be rotated away from the position for maximum reception more than is absolutely necessary, and in no case more than 45°.

Accuracy of Consol Bearings

The accuracy of Consol bearings is greatest on the perpendicular bisector of the line joining the aerials, that is the base line. Away from the base line the count lines and equisignal lines become increasingly farther apart, and the accuracy accordingly falls off as the angle from the normal to the base line increases.

The rotations of the dot and dash sectors on the two sides of the base line are in opposite directions to one another, and accordingly the wide sectors adjacent to the base-line extensions are such that the same signal is heard in adjacent sectors. Bearings obtained when the ship lies within any of these sectors are not only liable to great inaccuracy, but are liable to be interpreted wrongly. For these reasons the four sectors adjacent to the base-line extensions are not included in the coverage area of a Consol station. In setting up a Consol station the alignment of the aerials, if possible and practicable, is such that the ambiguous sectors adjacent to the base-line extensions lie over regions which normally have only a relatively low density of aircraft (or shipping).

The useful range of Consol along the normal to the base line has been estimated to be about 500 miles with a probable error in the position line of about 3 miles. At about 1,000 miles the probable error is about 6 miles. On the edge of the coverage area at these ranges, the probable position-line errors are about 12 and 24 miles respectively. At night, because of sky-wave radiation, accuracy is not so high. The estimated probable position-line error on the normal to the base line is about 10 miles at a range of about 1,000 miles, and about 40 miles at the same range on the edge of the coverage area.

It has been explained that a Consol position line is a line along which the difference of phase between the waves received from the two outer aerials is constant. Any factor which causes a change of phase of either wave will therefore cause error. Phase changes which will cause error may arise at the transmitting station; the receiving station; or during the passage of the radio energy from the Consol station to the receiver.

At the transmitting station phase changes may take place in the transmitter and aerial circuits, and may also arise through changes in the efficiency of radiation from the aerials. If the electrical conductivity of the ground or air in the vicinity of the aerials changes, because of changes in meteorological conditions, the phases of the radiated waves may be affected. Hills and buildings in the close vicinity of the Consol station may also cause errors through possible refraction or reflexion of the transmitted energy.

At the receiving station error may result through the use of an automatic volume control (A.V.C.), should one be incorporated in the set. The object of A.V.C. is to compensate for fading by keeping the amplitude, and therefore the signal strength, constant. The effect of A.V.C. when receiving Consol waves is to reduce the change in intensity of the signal when keying takes place, thus causing a broadening of the equisignal line. A loop aerial may also cause error. If the loop aerial or search coil is set to the position for minimum reception, large errors may result especially at short ranges. In the minimum reception position, the ground waves—which are horizontally polarized—do not induce current in the loop aerial, but sky waves, especially at short ranges, when they strike the loop aerial at relatively large angles to the horizontal, do cause induction. False counts may therefore arise if the range is such that the ground-wave signals are stronger than the sky-wave signals. Hence the reason, when using a direction finder receiver for Consol observations, to set the loop aerial or search coil to the position for minimum reception.

The largest errors which may affect Consol observations are brought about during the passage of the radio energy from the transmitter to the receiver. These errors are known as propagation errors. The ground radiation from a Consol station is affected by the conductivity of the surface of the earth over which it passes. If this varies along the path, a change of phase of the waves at the receiver may result in a shift of the equisignal line from its theoretical position, thus causing an error in the Consol position line.

When receiving ground- and sky-wave radiation simultaneously, errors may result due to interference between the two. The range at which sky waves from Consol stations become predominant is approximately 500 miles over the sea. At ranges of less than about 150 miles, the ground radiation alone is normally being received. Between these distances both ground waves and sky-waves are received, and within these ranges, therefore, errors due to interference may result. If the times taken for radiated energy, whether ground- or sky-wave, to travel from a Consol station to a receiver, from the two outer aerials, are different, a relative change of phase of the signals received takes place, with the possibility of error in the observed Consol count.

The operational range of Consol is limited by interference and static received by the receiver. Interference may result through transmitters—other than that of the Consol station—working at the same time on nearby frequencies. It may also result from the magnetic fields which emanate from badly screened radio and electrical appliances. Static is the name given to the energy received from electrical discharges which take place in thunderstorms. Static occurs on all radio frequencies, especially the lower frequencies, as the lower-frequency electric waves generated in thunderstorms are attenuated least and therefore travel the farthest.

The intensity of static received is dependent upon the meteorological conditions prevailing between the transmitter and the receiver. These conditions vary from hour to hour, day to day, and from season to season. They also vary with location on the Earth's surface.

The amount of static received is also a function of the bandwidth of the receiver. The bandwidth of a receiver decides the selectivity, this being the ability of the receiver to distinguish between signals transmitted on nearly equal frequencies. All receivers receive radio waves within a narrow band on both sides of the frequency to which the receiver has been tuned. The narrower the bandwidth, the larger will be the ratio between signal strength and static and noise. The term noise applies to the crackling sounds which are due to faults, often irremedial, in the receiver and aerial circuits. The limit of range of Consol signals is considered to be reached when the signal strength has fallen to twice the static and receiver noise.

Consol is an aid to navigation which is most useful for ocean navigation. Bearings are in general insufficiently accurate, because of the several possible causes of error, for making landfalls or for navigating close inshore. Mariners are therefore warned that Consol bearings should not be relied on implicitly when closing danger.

Postscript

From the mariner's viewpoint a radio direction finder may be considered to be a medium-range navigational aid, of greatest value in times of poor visibility when astronomical or visual methods of position fixing are not available. Bearing in mind the several errors which may affect radio bearings, the results of radio direction finding should always be treated with reserve and the necessary seamanlike caution. No prudent navigator would place implicit faith in fixes or position lines obtained by his radio direction finder, no more than he would place unquestioning trust in the results of any other radio navigational aid. However, with a well-maintained and properly calibrated direction finder; a skilful and intelligent observer; and propagation errors, such as those due to land effect and night effect, non-existent or of negligible magnitude, the probable error of a radio bearing may be regarded as being within about 2° of the truth.

In an analysis of D.F. bearings reported over a period of two years, as many as 87 per cent of about 2,000 radio bearings which were visually checked had an error of less than 1°, and only about 5 per cent had errors greater than 2°. It would be expected that the best results of radio direction finding are obtained on ships on which the instrument has been properly calibrated, is properly cared for, and is operated by knowledgeable and skilful observers, but obviously these factors could play no part in the analysis, brief details of which are given above.

After obtaining a radio D.F. bearing (and this applies to all observed bearings), the observer should estimate its degree of accuracy. Then, instead of laying down—as is customary—a position line, which may result in a false sense of accuracy, a sector, the angular width of which corresponds to the estimated degree of accuracy of the bearing, should be drawn. It will be remembered that an error in laying off a position line produces a displacement of the position line from the true position of the ship, about 1 mile for each degree of error per 60 miles of distance between the ship and the observed object or radio beacon. This rough rule holds good for errors up to about 5°. Thus, when choosing radio beacons for the purpose of fixing by radio bearings, the distance between the ship and the beacons chosen ought to be considered. In general, the beacons which have the smallest distances from the ship—providing that their bearings differ by more than about 30°—produce the best radio fixes. An error of 2° gives rise to a displacement of the position line from the true position of the ship of 4 miles, if the distance between the beacon and the ship is 120 miles. If the distance is 240 miles, the displacement for the same error is 8 miles. If the distance is only 15 miles, the displacement of the position line as a result of an error of 2° is only a half a mile.

By observing three radio bearings, and correcting them if necessary for half convergency, the resulting position lines will normally cross to produce a cocked hat. The ship's position is usually estimated to lie somewhere within this triangle. Now the smaller the distances between the ship and each of the three beacons, the smaller will be the resultant cocked hat, and the less ambiguous will be the position determined. Now although it is true that the shorter the distance between the ship and the radio beacon, the greater will be the absolute accuracy of a position obtained from its bearing, the practice of navigating solely by radio direction finder when close to the land, or for making harbour, is not recommended. Anyone so doing, except in extreme circumstances, must be regarded as imprudent, to say the least. It is necessary to use discretion when using the results of any aid to navigation.

To use radio bearings for fixing a ship's position, when the ship is many hundreds of miles from the beacons observed, should also be regarded as a navigational absurdity, in view of the fact that a small error in the observation produces, under such circumstances, large displacements of the position lines from the actual position of the ship.

The radio direction finder is most useful for position fixing when within the ground range of nearby radio beacons, and it is particularly valuable for making landfalls after an ocean passage during the latter part of which astronomical observations have not been possible.

In some coastal areas, the spatial density of radio beacons is high, as for example in North West Europe or North America. In such regions radio position lines may afford valuable information for the safe navigation of the ship. It has been stated—and doubtless it has been done—that it is possible to navigate through the English Channel by means of a radio direction finder alone, to an accuracy of about half a mile.

In parts of the world where the spatial density of radio beacons is low, use may be made of Coast Radio Stations, for the purpose of determining radio position lines. These stations normally operate on a frequency of 500 kc/s, and it is usual, therefore, to have the ship's radio direction finder calibrated for this frequency as well as for the radio beacon frequency.

Besides the importance of a radio direction finder for position fixing, it may be used for the purposes of homing on a transmitting station. This is particularly useful when the transmitting station is a ship or aircraft in distress, or a lifeboat or raft from a wrecked vessel. The use of a direction finder for homing on a radio beacon sited on a light vessel, especially during low visibility, is the subject of a warning to mariners. The danger of so doing was exemplified by a regrettable disaster when the Nantucket light vessel was rammed, in fog, by a ship which was homing on the light vessel by means of her radio direction finder.

The radio direction finder may be used for receiving radio messages which cannot be received, on account of jamming or excessive interference, by means of a receiver with an omnidirectional aerial. It may also be used for identifying ships in an anchorage or harbour approach, when fog renders it impossible to sight them. In this respect it may be regarded as a useful auxiliary instrument to radar.

Following World War 2, during which intensive radio research was conducted because of the need for improved methods of position fixing and detection of ships, several new radio aids to navigation were introduced to the mariner. The radio direction finder can now, therefore, no longer claim the great relative importance which it could justifiably claim during the period from its inception in 1908 to 1939 when war broke out. The opinion of Sir Robert Watson-Watt, published in the first volume of the *Journal of the Institute of Navigation*, is interesting. "The radio direction finder," he wrote, "is a device of rapidly diminishing usefulness, which is quite worth retaining in service for some years to come, but which should not be regarded as being still, at this date (1948), that major contribution from radio to navigation which it was in its early days."

Despite its fall in relative importance since the introduction of radar aids to navigation, the radio direction finder will doubtless still rank for some time yet as an instrument of value to navigators. The representatives of the principal maritime nations, at the International Meeting on Radio Aids to Navigation (I.M.R.A.N.), concluded that medium-frequency direction finding can continue to fulfil a useful function in making landfall and in coastal navigation, and for search and rescue requirements.

In conclusion, the remarks made by the author (F. P. Best) of a paper on marine direction finding, read before the Institute of Navigation, are relevant. "Radio direction finding," it was claimed, "is the oldest, simplest, and probably most reliable radio aid yet offered to the mariner. In some ways it may be likened to a magnetic compass, and, if its range is very much smaller, it has the inestimable advantage over a compass of giving bearings on a very large number of stations instead of only one. It is the only navigational aid in which the performance is controlled by the shipborne installation, and is not subject to technical failures which may occur in land-based transmitters."

MINISTRY OF TRANSPORT AND CIVIL AVIATION

NOTICE No. M.402

INSTALLATION, CALIBRATION, AND MAINTENANCE OF DIRECTION FINDERS

NOTICE TO SHIPOWNERS, MASTERS, AND MARINE RADIO COMPANIES, AND TO OWNERS AND SKIPPERS OF FISHING VESSELS

(This Notice replaces Notice No. M.372)

The attention of all concerned with the fitting and use of direction finders on ships of any kind is drawn to the necessity of ensuring that such equipment is properly installed, maintained, and fully and accurately calibrated.

Part I Installation

1. The loop aerial system should be mounted as near as practicable to the centre line of the ship, and as far as possible from the aerials of other radio equipment, and from large movable metal objects such as derricks and wire halyards.

2. Fixed metallic masses may give rise to considerable errors depending on their distances from the loop aerial system. The latter should be placed not less than 6 ft from such masses which rise above the base of the loop. The best results are obtained when the distribution of the metallic masses is symmetrical in relation to the loop.

3. Wire stays should be broken by insulators if they are so close to the loop aerial that the accuracy of the direction finder is likely to be prejudiced. Long wire stays should be broken into lengths not exceeding 20 ft.

4. It should be ensured that cable joints are watertight, particular attention being given to joints where the cable enters the deck-head of the space in which the direction finder is installed. On the other hand arrangements may be necessary at the foot of the pedestal to permit egress of condensed water and to permit suitable ventilation of the pedestal and loop system.

5. Unless the feeders between the aerial system and the receiver are of solid-dielectric screened (coaxial) cable they should be protected by metal tubes.

6. The loop aerial pedestal and any protecting metal tubes for feeders should be bonded to earth.

7. The sense aerial should be as short as is compatible with effective sense-finding.

8. The apparatus should be free from interference, either mechanical or electrical, which would be likely to hinder the proper reception of signals for determining bearings. Interference arising from induction caused by fans, motors, ships' mains, and other electrical sources should be carefully

watched, and, where necessary, remedial measures such as the fitting of filters or suppressors should be introduced. British Standards Specification No. 1597 of 1949, which deals with radio interference suppression, lays down permissible limits of interfering voltages, measures to be taken in connexion with radio receiving installations, aerials, rigging, electrical wiring, and electrical machinery, and the standards of components recommended for suppression purposes. The practices recommended in this specification should be followed as far as practicable, including the use of components which meet the standard specified.

9. An efficient two-way means of calling and voice communication should be provided between the direction finder receiver position and the bridge from which the ship is navigated. There should also be an efficient means of signalling between the direction finder receiver position and the ship's standard compass or gyro compass repeater for use when calibrating or taking check-bearings.

10. A schematic wiring diagram and book of instructions should be supplied with the apparatus and kept readily available on board.

Part II Calibration

11. The direction finder should be calibrated after installation in the ship, whenever the position of the loop aerial system is changed, or whenever any alteration is made to any structure or fitting on deck which is likely to affect the accuracy of the equipment. In each instance calibration should be completed either before the commencement or in the early stages of the first voyage using the first calibrating facilities which become available.

12. Before calibration starts the apparatus should be carefully checked for mechanical and electrical performance.

13. (1) Other aerials in the ship can cause serious errors to radio bearings especially when above or near the loop aerial and when connected to transmitters or receivers which are tuned to or near the frequency on which radio bearings are being taken.

(2) All aerials which have any part rising above the base of the loop aerial, within 50 ft horizontal distance of it, should be kept isolated during calibration and whenever the direction finder is actually in use unless and until they have been individually tested and found not to introduce any serious error.

(3) Such tests should be made on first installation by noting the maximum effect on radio bearings when the aerial is alternately connected to and isolated from equipment tuned to the frequency on which the test radio bearings are being taken or as near thereto as its design permits.

(4) The aerials used for maintaining the safety watch, i.e. those to which the emergency receiver and/or the auto-alarm are normally connected should, where practicable, be erected outside the 50 ft radius mentioned above. If this is impracticable the aerial or aerials should be tested as in sub-paragraph (3) above, and if found to cause serious error to radio bearings should be kept isolated during calibration and whenever the direction finder is actually in use on the frequencies on which these errors have been found. In these circumstances, care should be taken to ensure that the safety watch is disturbed for as short a time as possible.

(5) The haphazard erection of broadcast aerials can be a serious source of error in radio bearings. It is strongly recommended that all broadcast receivers should be attached to either

(*a*) a single communal aerial;
(*b*) aerials which do not rise above the base of the loop aerial; or
(*c*) aerials which are outside the 50 ft radius mentioned in sub-paragraph (2).

Any broadcast aerial, other than those mentioned in (*b*) and (*c*) above, should always be dealt with in the manner described in sub-paragraphs (2) and (3).

(6) The calibration data should include a list of all aerials showing their condition (i.e. whether isolated or connected) and also the position and condition of any movable deck structures, etc. (see paragraphs 1 and 2) so that the same conditions may obtain when the direction finder is used navigationally.

(7) Any additional aerial coming within the above conditions as to height and distance from the loop aerial should be tested immediately as described and its particulars and condition for use entered in the record.

14. Calibration should be effected by two persons, one experienced in the taking of radio bearings and the other in the taking of visual bearings.

15. Calibration should be carried out clear of the land. It may be effected by either

(*a*) swinging the ship in relation to a fixed radio station, or
(*b*) receiving signals transmitted by an auxiliary ship while circling the stationary ship.

16. When possible the fixed radio stations referred to in paragraph 15(*a*) should be a special calibration station or a maritime radio beacon transmitting a suitable and adequately strong signal. The siting conditions of other radio stations may affect the accuracy of bearings. See also paragraph 20.

17. The ship should be so placed that throughout the period of calibration the position of the aerial of the calibrating transmitter can be seen. The distance between the calibrating transmitter and the direction finder should not be less than 1 mile; the ship should not be less than this distance from land.

18. Visual and radio bearings should be taken simultaneously at intervals of not more than 5° in all arcs in which visual bearings are possible. The visual observer should stand as near as possible to the loop aerial as otherwise a parallax error may enter the calculations.

The radio bearings should be taken on a frequency within the band 285–315 kc/s (radio beacon band). In each instance the visual and radio bearings and the difference between them should be recorded.

19. Calibration tables and curves should be prepared from the details of visual and radio bearings taken during the calibration period. These tables and curves and data as required by paragraph 13(6) should be kept on board for the use of any person operating the direction finder. Radio bearings when corrected from the curves should not differ from the correct bearings by more than 2°.

20. With reference to paragraphs 15 and 16 it should be noted that unless the transmitting ship or station uses either—

(1) a vertical transmitting aerial, or
(2) a "T" aerial in which the horizontal limbs are substantially symmetrical with respect to the vertical limb;

large and variable errors in bearing may be experienced.

21. Details of established calibration facilities will be found in *Admiralty List of Radio Signals,* Vol. II.

Part III Check-bearings

22. Check-bearings should be taken so that the tables and curves are fully verified in each period of twelve months, or whenever any changes are made in any structure or fitting on deck which is likely to affect the accuracy of the equipment. During these checks, the effect on radio bearings of aerials falling within the limits detailed at paragraph 13(2) should be confirmed.

23. When at sea, check-bearings should be taken as often as practicable on all frequencies liable to be used for the taking of navigational bearings. The usual method of checking the accuracy of calibration data is to take a series of radio bearings and visual bearings while the ship is on passage in the vicinity of a radio beacon or coast station. Provided the conditions stated in Part II are observed a good check is obtained of the performance of the equipment on the sector and at the frequency concerned. If the check-bearings disclose errors which differ from those indicated by the calibration curve, it should be remembered that these curves are not always purely quandrantal in shape; the new readings, therefore, should not be used to amend the calibration curve in unchecked sectors although they may be useful as a general guide. Useful check-bearings may also be taken on other ships at sea. When taking check-bearings on frequencies above 315 kc/s, it should be borne in mind that the calibration curve may not be accurate for such frequencies.

24. If at any time check-bearings show serious discrepancies from the data obtained at the last calibration, the direction finder should be used with extreme caution and should be fully recalibrated at the earliest opportunity.

Part IV Radio Navigational Bearings

25. When the direction finder is used for taking radio bearings for navigational purposes it is important that the movable deck structures and aerials of the other radio equipment in the ship should be in the same position and condition as when the direction finder was calibrated (*see* paragraph 13).

Part V Maintenance

26. A suitable and adequate supply of power should always be available for meeting the requirements of the direction finder on the particular ship in which it is fitted.

Where the emergency battery is used for providing current to the direction finder the radio officer should test the battery once a day by voltmeter and once a month by hydrometer, and keep it in such condition that it can at any time provide the supply of power which must always be available for radio emergency and other purposes. If the direction finder is powered from some other battery, this battery should be similarly tested and maintained.

27. Care should be taken that any insulating segment of the loop aerial is not painted over.

Part VI Compulsorily Equipped Ships

28. In ships which are required by the Merchant Shipping (Direction Finders) Rules, 1952, to be provided with a direction finder the following records must be kept on board the ship in a place accessible to any person operating the direction finder, and available for inspection by the Radio Surveyor—

(1) a list or diagram indicating the position and condition at the last occasion on which the direction finder was calibrated of the aerials on board the ship and of all movable structures which might affect the accuracy of the direction finder (see paragraph 13);

(2) the calibration tables and curves which were prepared as a result of the last calibration (see paragraphs 19 and 22);

(3) a certificate of calibration in the form specified in the Rules signed by the persons who made the last calibration (see paragraph 14); and

(4) a record, in the form specified in the Rules, of all check-bearings taken for verification purposes, and of any other information which might be useful in showing whether or not the equipment gives satisfactory performance (see paragraphs 19 and 22). The bearings must be entered and numbered in the order in which they are taken.

(5) where applicable, a record of the battery tests (see paragraph 26).

The forms referred to in sub-paragraphs (3) and (4) above are reproduced as Appendices I and II of this Notice.

Ministry of Transport and Civil Aviation,
London.
December, 1956.
MNA. 33/8/016

APPENDIX I

CERTIFICATE OF CALIBRATION OF DIRECTION FINDER

We, the undersigned, hereby certify that we have this day

(*a*) calibrated in accordance with the Merchant Shipping (Direction Finders) Rules, 1952, the direction finder installed in the

$\frac{\text{s.s.}}{\text{m.v.}}$ ....

(*b*) handed to the Master of that ship tables of calibration corrections.

(*c*) adjusted the said direction finder so that the readings taken thereby, when corrected with such tables, differed from the correct bearings by no more than plus or minus two degrees.

We hereby further certify that the Master of the said ship has been furnished with a list or diagram indicating the conditions and position, at the time of such calibration, of the aerials and of all movable structures on board the ship which might affect the accuracy of the direction finder.

....................................Radio Observer

....................................Visual Observer

........................Date

APPENDIX II

RECORD OF CHECK-BEARINGS TAKEN BY MEANS OF THE DIRECTION FINDER

Serial Number of Bearings	Date	Time (G.M.T.)	Ship's Approximate Position		Distance from Transmitter	Direction Finder Bearing of (Name)	Direction Finder Relative Bearing Corrected for Q.E.	Ship's Head by Compass 0/360°	Total Compass Error	½ Convergency Applied	Ship's Head Corrected (True)	True Bearing by Direction Finder (Col. (8) and Col. (12))	True Bearing by Calculation or by Visual Check (whether calculated or visual to be indicated)	Correction Required to Make Col. (13) Equal Col. (14) (Indicating Whether — or +)	Signature of Observer or Observers
			Latitude	Longitude											
(1)	(2)	(3)	(4)	(5)	(6)	(7)	(8)	(9)	(10)	(11)	(12)	(13)	(14)	(15)	(16)

Bibliography

The Admiralty List of Radio Signals, Volume 2 (1959).

The Admiralty List of Radio Signals, Volume 5 (1954).

Handbook of Wireless Telegraphy, H.M.S.O. (1938).

Wireless Direction Finding, R. Keen (1949).

Navigational Wireless, S. H. Long (1927).

Handbook of Technical Instruction for Wireless Telegraphists, H. M. Dowsett and L. E. Q. Walker.

The Calibration of Direction Finders, F. P. Best and J. H. Moon.

An Introduction to Direction Finding and Navigation, L. Bainbridge-Bell.

Journals of the Institute of Navigation.

INDEX

CPSIA information can be obtained
at www.ICGtesting.com
Printed in the USA
BVHW052241090223
658263BV00003B/88